Analysis and Mathematical Models of Canned Electrical Machine Drives

Qiang Yu • Xuesong Wang • Yuhu Cheng
Lisi Tian

Analysis and Mathematical Models of Canned Electrical Machine Drives

In Particular a Canned Switched Reluctance Machine

Springer

Qiang Yu
School of Electrical and Power Engineering
China University of Mining and Technology
Xuzhou, Jiangsu, China

Xuesong Wang
School of Information and Control Engineering
China University of Mining and Technology
Xuzhou, Jiangsu, China

Yuhu Cheng
School of Information and Control Engineering
China University of Mining and Technology
Xuzhou, Jiangsu, China

Lisi Tian
School of Electrical and Power Engineering
China University of Mining and Technology
Xuzhou, Jiangsu, China

ISBN 978-981-13-2744-5 ISBN 978-981-13-2745-2 (eBook)
https://doi.org/10.1007/978-981-13-2745-2

Library of Congress Control Number: 2018957462

This Springer imprint is published by the registered company Springer Nature Singapore Pte Ltd.
The registered company address is: 152 Beach Road, #21-01/04 Gateway East, Singapore 189721, Singapore

Preface

This book deals with analysis and mathematical models of an electrical machine, especially a canned machine as a hydraulic pump drive. The characteristic use of metallic can shields in air gap makes the electromagnetic and thermal features substantially different. The can effect is modeled and analyzed mainly based on a switched reluctance machine. Such work is about our 10 years' research on canned electrical machines.

The organization of this book is as follows. In Chap. 1, the canned electrical machine drives, recent development, state-of-the-art prototypes, and applications are introduced. In Chap. 2, electromagnetic analysis of saliency effect is studied, and flux-linkage characteristics are discussed for a canned switched reluctance machine. In Chap. 3, the can effect and operation principle of a novel canned switched reluctance machine are systematically illustrated. In Chap. 4, spatial distribution of magnetomotive force is modeled, while further in Chap. 5, an analytical model of concentric layer structure is developed for canned electrical machines. In Chap. 6, a novel thermal network model is proposed to analyze characteristic thermal features of a canned machine, and further electrothermal coupled analysis on cans is discussed.

This book works for liquid pump industry, for enterprises or institutions preparing to apply this technology, and for the field of electrical machine research and analysis. The target audience is located in electrical engineers, postgraduates, and technology researchers.

Xuzhou, Jiangsu, China Qiang Yu

Acknowledgments

This book is supported by the National Natural Science Foundation of China (NSFC grant no. 51607180). This work, in part, was originated and studied from Institute of Electrical Machines and Actuators (Elektrische Antriebstechnik und Aktorik, EAA), University of Bundeswehr Muenchen, Munich, Germany, where the first author of this book was pursuing his Ph.D. study during 2008–2012. The authors greatly acknowledge Professor Dr.-Ing. Dieter Gerling for his instructions and guidance during the research progress, Dr.-Ing. Gurakuq Dajaku and Dr. Christian Laudensack for their technical support, as well as KSB Aktiengesellschaft, Frankenthal, Germany, for industrial cooperation and application. The first author also greatly acknowledges the McMaster Automotive Resource Center (MARC), McMaster University, Hamilton, Ontario, Canada, where his postdoctoral research was carried out. The authors greatly acknowledge Professor Dr. Ali Emadi and Dr. Berker Bilgin for their technical support.

Finally, the authors would like to record their thanks to Publishing Editor, Jasmine Dou, for text reading and suggestions.

Acknowledgments

Contents

1 Overview of Canned Electrical Machines 1
1.1 Background 1
1.2 Research Overview 3
1.2.1 Structural Topology 4
1.2.2 Loss Analysis 5
1.2.3 Thermal and Acoustic Analysis 6
1.2.4 Review Summary 7
1.3 Canned Switched Reluctance Machines 7
1.4 Outline Summary 8
References 10

2 Electromagnetic Analysis of Saliency and Can Effect by Network Models 13
2.1 Flux Linkage Modeling of Switched Reluctance Machines 13
2.2 A Discretized Circuit Network Model 16
2.2.1 Modeling of Airgap Reluctance 16
2.2.2 Modeling of Pole Reluctance 18
2.2.3 Modeling of the End Part 20
2.2.4 Calculation Flow 20
2.2.5 Application Examples 21
2.2.6 Can Loss Analysis 24
2.3 Loss and Efficiency Analysis 29
2.3.1 The Calculation Method 29
2.3.2 The Variable Loss Coefficients 35
2.3.3 The Discretized Elements 37
2.3.4 Verification and Discussion 39
2.4 A Simplified Network Model 42
2.4.1 The MEC-FE Model 42
2.5 A Fitting Method for Airgap Reluctance 49

2.6 Chapter Summary 53
References 54

3 Electromagnetic Analysis of Can Effect of a Canned SRM 57
3.1 Canned Switched Reluctance Machine and Operation Principles . . 57
3.2 Eddy Current and Loss Features at Typical Rotor Positions 61
3.2.1 Single Phase Excitation 61
3.2.2 All-Phase Excitation 62
3.3 Can Loss Variation of One Stroke Period 63
3.4 Airgap Flux and Eddy Current Loss Due to the Use of Cans 66
3.5 Experimental Validation 67
Reference 70

4 An Analytical Model of Concentric Layer Structure for Canned Machines, Part I: Armature Coils 71
4.1 Model Introduction 71
4.2 Modeling of Winding Function 72
4.2.1 Fourier Approach for a Single Turn of Wire 72
4.2.2 Model of a Tooth Concentrated Coil with Wire Layout (Model 1) 76
4.2.3 Model of One Coil in Distributed Topology 86
4.2.4 Model of Coil Distribution 88
4.3 Modeling of Phase Current 92
4.4 Modeling of MMF Distribution 94
4.4.1 Concentrated Coils 94
4.4.2 Distributed Coils 96
4.5 Simulation and Discussion 100
4.6 MMF Distribution from Axial Direction 103
4.7 Application Example 104
4.8 Modeling of Permanent Magnets 106
4.9 Chapter Summary 111
References 112

5 An Analytical Model of Concentric Layer Structure for Canned Machines, Part II: Magnetic Field 113
5.1 Model Introduction 113
5.2 The Model of Concentric Layer Structure 116
5.2.1 Vector Potential of the First Layer 119
5.2.2 Vector Potential of More Layers 121
5.2.3 Coordinate Transformation 121
5.2.4 Calculation of Constants 123
5.3 Magnetic Field due to DC Potential 126
5.4 Modeling of Saliency Effect 129
5.5 Feature Deduction 131
5.5.1 Flux Density of Stator 131
5.5.2 Flux Linkage 132

5.5.3 Iron Loss 134
5.5.4 Can Loss 136
5.5.5 Torque 137
5.6 Verification and Discussion 140
5.6.1 Canned Induction Machine 140
5.6.2 Canned Permanent Magnet Machine 146
5.6.3 Canned Switched Reluctance Machine 152
5.7 Chapter Summary 154
References 155

6 Thermal Analysis of a Canned SRM 157
6.1 Background 157
6.2 The Thermal Network Model 160
6.2.1 Model Overview 160
6.2.2 Heat Sources 161
6.2.3 Thermal Resistances 161
6.2.4 Model of Coils 167
6.3 Application and Discussion 173
6.4 Electro-thermal Coupled Analysis 180
6.4.1 The Calculation Flow 180
6.4.2 Application and Discussion 182
6.5 Chapter Summary 183
References 184

7 Conclusions and Future Work 185
7.1 Conclusions 185
7.2 Future Work 187

Appendix 189

About the Authors

Qiang Yu received his Ph.D. degree from the University of Bundeswehr Muenchen, Munich, Germany, in 2012. From 2008 to 2012, he was an engineer in FEAAM GmbH, Neubiberg, Germany. From 2013 to 2014, he was a postdoctoral research associate at McMaster University, Hamilton, Ontario, Canada. From 2014 to 2015, he was a postdoctoral research fellow in Université Libre de Bruxelles, Brussels, Belgium. Currently, he is an associate professor in School of Electrical and Power Engineering, China University of Mining and Technology. His research interests include modeling, design, and control of electrical drives and systems.

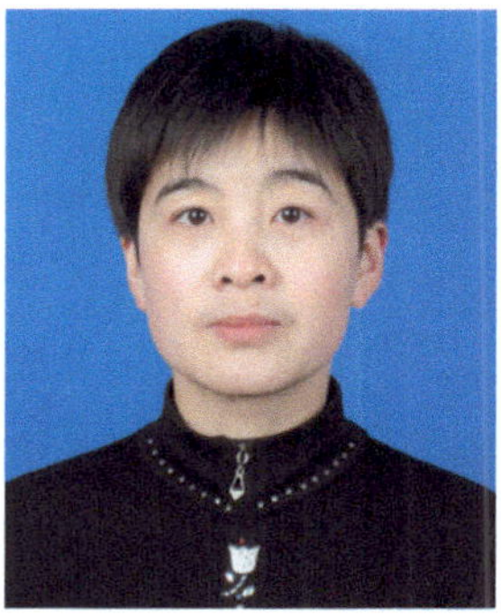

Xuesong Wang received her Ph.D. degree from China University of Mining and Technology in 2002. She is currently a professor in School of Information and Control Engineering, China University of Mining and Technology. Her main research interests include electrical drives, bioinformatics, and artificial intelligence. In 2008, she was recipient of the New Century Excellent Talents in University from the Ministry of Education of China.

Yuhu Cheng received his Ph.D. degree from the Institute of Automation, Chinese Academy of Sciences, in 2005. He is currently a professor in School of Information and Control Engineering, China University of Mining and Technology. His main research interests include electrical drives and intelligent systems. In 2010, he was recipient of the New Century Excellent Talents in University from the Ministry of Education of China.

Lisi Tian received his Ph.D. degree from Huazhong University of Science and Technology (HUST), China, in 2015. He is currently with School of Electrical and Power Engineering, China University of Mining and Technology. His main research interests include power electronics, electrical drives, and fault diagnosis.

List of Figures

Fig. 1.1	Comparison of the traditional and improved hydraulic pump systems	2
Fig. 1.2	The integrated system and the canned drive machine	2
Fig. 2.1	Definition of airgap magnetic flux paths when poles don't have degree of alignment	16
Fig. 2.2	The relation between physical distance of pole tips AD (mm) and modeled width of the main flux path w_{p1} (mm) when poles don't have alignment as depicted in Fig. 2.1, showing independence of w_{p1} to machine geometry variation	17
Fig. 2.3	Definition of airgap magnetic flux paths when poles have degree of alignment	18
Fig. 2.4	Discretizing pole reluctances according to airgap paths, (**a**) partial overlap, (**b**) non-overlap	19
Fig. 2.5	Determination of pole tip reluctances that linked with the main airgap flux path at non-overlap and overlap rotor positions	19
Fig. 2.6	Flux lines at the end part, (**a**) fully aligned, (**b**) partial aligned, (**c**) non-overlapping	20
Fig. 2.7	The calculation flow with a dual-loop structure	21
Fig. 2.8	The flux paths and airgap permeance variation of a pole pair by FE and the proposed analytical network. Note that in FE, electrical steel does not saturate. Main geometrical variation of the machine includes (**a**) the original machine, (**b**) with increased rotor pole width M1, (**c**) with enlarged airgap length M2, (**d**) with both increased stator and rotor pole widths M3	22

Fig. 2.9 Flux linkage characteristics of machines in Table 2.2, (**a**) original M0, in which A = aligned, B = partial overlap, C = just after overlap, D = just before overlap, E = unaligned, (**b**) increased rotor pole width M4, in which A = aligned, C = just after the start of overlap, D = just before the start of overlap, E = unaligned, (**c**) reduced airgap length M5, in which A = aligned, B = partial overlap, C = just after the start of overlap, D = just before the start of overlap, E = unaligned. Dots are FE data while curves are analytical data 23
Fig. 2.10 Phase torque curves of machines in Table 2.2, with phase current variation, (**a**) original M0, (**b**) increased rotor pole width M1, (**c**) reduced airgap length M2. In all cases, phase current levels (in ampere) are, A = 30, B = 60, C = 90 and D = 120. Dots are FE data while curves are analytical data 25
Fig. 2.11 The network update by adding rotor and stator cans, and note that the radial length values of airgap and cans are disproportionally enlarged 26
Fig. 2.12 FE modeling and analysis of a canned switched reluctance machine, (**a**) Geometrical overview with can shields highlighted, (**b**) detailed flux paths and density distribution 26
Fig. 2.13 The calculation flow 27
Fig. 2.14 Comparison of flux linkage characteristics of a canned machine 28
Fig. 2.15 Flux linkage curves and can effect comparison at selected rotor positons 29
Fig. 2.16 Flux density variation of a single element on the stator and rotor teeth on different excitation modes for the studied 4-phase 16/12 traction SRM (**a**) low speed (**b**) high speed 30
Fig. 2.17 The flux reversals by the fundamental and harmonics 31
Fig. 2.18 The flux reversals by the fundamental and harmonics 31
Fig. 2.19 The flux density waveform of an element showing the local flux reversals 33
Fig. 2.20 Separation of reversals into main and minor loops 33
Fig. 2.21 Hysteresis loss by different methods 34
Fig. 2.22 The loss curves of M1529G 35
Fig. 2.23 Linearized loss curves 36
Fig. 2.24 Loss coefficients with flux density, (**a**) hysteresis, (**b**) eddy 36
Fig. 2.25 The magnetic equivalent paths when teeth are not aligned 37
Fig. 2.26 The flow chart to get all elements in the proposed method 38
Fig. 2.27 The phase inductance characteristics showing variation of the air gap plus saturation using the proposed magnetic circuit method 38
Fig. 2.28 The flow chart of the program 40
Fig. 2.29 The current-loss relationship from different methods at low speed 40

Fig. 2.30 The speed loss relationship from different methods ... 41
Fig. 2.31 Comparison between measured and predicted losses under no saturation condition (constant torque 5 Nm) ... 41
Fig. 2.32 Comparison between measured and predicted losses with saturation, with constant low speed 1000 rpm ... 42
Fig. 2.33 Flux paths overview at typical rotor positions, (**a**) at unaligned position by one single phase excitation, (**b**) at partial aligned position by two phase excitations, showing magnetic coupling effect ... 43
Fig. 2.34 Overview of the proposed network model ... 44
Fig. 2.35 Airgap flux path details, showing the non-overlap and partial overlap of a pole pair ... 44
Fig. 2.36 The permeance drop lookup function obtained from FE, (**a**) non-overlapping, (**b**) partial and full overlapping ... 45
Fig. 2.37 The flux linkage under selected rotor positions without magnetic coupling ... 45
Fig. 2.38 Comparison of flux linkage curves under selected rotor positions, with and without magnetic coupling between phases ... 47
Fig. 2.39 Comparison of flux linkage variation due to the coupling effect, where different levels of current excitation are applied ... 47
Fig. 2.40 The magnetic coupling effect due to variation of airgap length, where M1 and M2 are simulated, with M0 as reference; Note that 22.5° is the aligned position and phase current 170A is applied; Dots: with phase coupling; Lines: no phase coupling ... 48
Fig. 2.41 The magnetic coupling due to variation of pole widths where M3 and M4 are simulated, with M0 as reference; Note that 22.5° is the aligned position and phase current 170A is applied; Dots: with phase coupling; Lines: no phase coupling ... 49
Fig. 2.42 The magnetic coupling effect due to stator yoke widths where M5 is simulated, with M0 as reference; Note that 22.5° is the aligned position and phase current 170A is applied; Dots: with phase coupling; Lines: no phase coupling ... 49
Fig. 2.43 The proposed MEC network with a fitting method for airgap reluctance ... 50
Fig. 2.44 Geometrical definition of positions for a pole pair, (**a**) non-overlap, (**b**) partial overlap ... 50
Fig. 2.45 The phase inductance-position relationship by the proposed fitting model ... 52
Fig. 2.46 The magnetic curves and phase torque of the machine M0 in Table 2.10 ... 52
Fig. 2.47 The magnetic curves and phase torque of the machine M1 in Table 2.10 ... 53

Fig. 2.48 The magnetic curves and phase torque of the machine M2 in Table 2.10 53

Fig. 3.1 Sketch of the studied canned SRM 58
Fig. 3.2 Phase current with rated hysteresis band 60A, (**a**) fixed control mode at speed 2000 rpm, (**b**) turn-on advancing control mode at 3000 rpm, (**c**) continuous excitation control mode at 4500 rpm 59
Fig. 3.3 Definition of typical rotor positions of phase fluxing and defluxing 61
Fig. 3.4 Eddy current circulation on stator can by single phase excitation, where phase current corresponds to Fig. 3.2a; (**a**) The secondary circulation at the moment of fluxing; (**b**) The principal circulation at the moment of defluxing. The rectangular zone indicates the joint of the can and the flux/defluxing stator pole. Note that the numerical scale in (**b**) is more than twice of (**a**) 62
Fig. 3.5 Geometric overview of the canned machine and distribution of eddy current loss density on rotor/stator, showing typical loss generation due to phase fluxing and defluxing simultaneously; Note the definition of X-Y-Z coordinate, where Z is the axial direction 63
Fig. 3.6 The eddy current distribution on the can shield, at the rotor position corresponding to Fig. 3.5, showing the overlapped current circulations by both phase fluxing and defluxing; Note that the spatial position of the X-Y-Z coordinate corresponds to 90° clockwise rotation along Y-axis from the coordinate in Fig. 3.5 63
Fig. 3.7 The can loss variation at low speed 64
Fig. 3.8 Eddy current distribution on the stator can when the can loss is undergoing the principal peak, with stator poles and in particular the defluxing one shown as reference, (**a**) at the first sector with phase defluxing; (**b**) at the second sector with residual flux; The phrase "stator pole" means the joint zone of the stator can and the defluxing stator pole. Note that the numerical scale in (**a**) is twice of (**b**) 65
Fig. 3.9 Loss variation of the stator can at medium and high speed 65
Fig. 3.10 Airgap flux distribution of a standard and a canned SRM at the typical rotor position corresponding to Fig. 3.5 66
Fig. 3.11 The testing system setup 67
Fig. 3.12 The can loss variation with speed, under phase current levels measured by percent of the rated hysteresis limit 60A, *S* simulation, *M* measurement 69

Fig. 4.1 Arbitrary layout of a single wire (**a**) and corresponding winding function (**b**) 72

Fig. 4.2 MMF distribution of a single wire under ideal condition 73
Fig. 4.3 Improved models of a single turn, (**a**) concentrated, (**b**) distributed 75
Fig. 4.4 Definition of concentrated armature coils 76
Fig. 4.5 Geometrical specifications of a single turn for tooth concentrated coils 75
Fig. 4.6 The wire thickness factor variation by circumferential harmonic ν 78
Fig. 4.7 The short pitch factor variation by circumferential harmonic ν 79
Fig. 4.8 Wire distribution overview and definition of a layer, (**a**) structural overview, (**b**) regrouping 79
Fig. 4.9 Definition of a layer of wires within a coil, (**a**) $S = 2$, (**b**) $S = 3$ 80
Fig. 4.10 Transfer a coil onto slot surface, (**a**) a slot and coil, (**b**) equivalent transfer 84
Fig. 4.11 Modeling of tooth concentrated coils, (**a**) coil distribution, (**b**) equivalent transfer 85
Fig. 4.12 The winding function curve of a tooth concentrated coil using a simplified approach 85
Fig. 4.13 Spatial distribution of all coils 89
Fig. 4.14 Typical distributed winding topology of an AC machine 90
Fig. 4.15 Model of phase current 92
Fig. 4.16 Model of the rates of change of phase current determined by θ_1 and θ_2 93
Fig. 4.17 Phase current harmonics by different rates in Fig. 4.16, (**a**) $\theta_2 = 0.2$ rad, (**b**) $\theta_2 = 0.24$ rad, (**c**) $\theta_2 = 0.28$ rad 94
Fig. 4.18 The winding factor harmonics with different wire geometries, (**a**) Model 1, $d_w = 0.003$ m, (**b**) Model 1, $d_w = 0.002$ m, (**c**) Model 2 101
Fig. 4.19 The winding function distribution, (**a**) Model 1, (**b**) Model 2 102
Fig. 4.20 MMF harmonics at different time 102
Fig. 4.21 Phase current waveforms and selected simulation points A_1–A_6 103
Fig. 4.22 MMF distribution at selected rotor positions 104
Fig. 4.23 Windings and current flow in axial Z- direction 105
Fig. 4.24 Model of MMF distribution in axial Z-direction 105
Fig. 4.25 The MMF distribution in axial direction 105
Fig. 4.26 Flux paths overview at typical rotor positions, (**a**) at unaligned position by one single phase excitation, (**b**) at partial aligned position by two phase excitations, showing coupling effect 107
Fig. 4.27 Airgap flux density waveform when $S = 2$, each rotor position corresponds to Fig. 4.26, (**a**) one phase is excited at unaligned position, (**b**) two phases are excited at the same time 107

Fig. 4.28 Flux density distribution with different layers of wires $S = 1,2,3$, and each rotor position corresponds to Fig. 4.26, (**a**) one phase is excited at unaligned position, (**b**) two phases are excited at the same time ... 108
Fig. 4.29 Flux density harmonics with different layers of wires inside a coil ... 108
Fig. 4.30 Modeling of magnetization, (**a**) radial, (**b**) diametrical ... 109
Fig. 4.31 Magnetization in axial direction ... 111

Fig. 5.1 The proposed concentric multiple-layer structure model of a canned induction machine, (**a**) cylindrical components and settings, (**b**) rotor details and the coordinate definition ... 114
Fig. 5.2 The concentric multiple-layer structure model of a canned PM machine, (**a**) overview, (**b**) rotor details with PMs and the rotor can, with cylindrical coordinate defined ... 115
Fig. 5.3 The concentric multiple-layer structure model of a canned switched reluctance machine, (**a**) overview, (**b**) rotor details, the stator can and coordinate definition ... 115
Fig. 5.4 Boundary connection between inner surface of stator bore and outer surface of the first layer ... 117
Fig. 5.5 Geometrical definition of each layer ... 130
Fig. 5.6 Comparison of flux paths of the ordinary and canned machine under the same condition ... 141
Fig. 5.7 Flux density variation with rotor speed by the proposed model, (**a**) ordinary machine, (**b**) canned machine ... 142
Fig. 5.8 Can effect in terms of iron loss, (**a**) ordinary machine, (**b**) canned machine ... 142
Fig. 5.9 The normal flux density distribution of cans under different operation condition, (**a** and **b**), at synchronous speed, (**c** and **d**), at rated speed, (**e** and **f**), at zero speed, (**a**, **c** and **e**), stator can, (**b**, **d** and **f**) rotor can ... 143
Fig. 5.10 The normal flux density harmonics of cans at different condition, (**a**) stator can, (**b**) rotor can. Note: 1-synchronous speed, 2-rated speed, 3-zero speed ... 143
Fig. 5.11 Schematic of the test bench, showing the dSPACE 1103 system ... 145
Fig. 5.12 The total loss from the analytical method and experiment ... 145
Fig. 5.13 Schematic diagram of a canned SPM machine ... 147
Fig. 5.14 Flux path and density distribution due to cans, (**a**) ordinary machine, (**b**) canned machine ... 148
Fig. 5.15 Airgap flux density harmonics at different occasions of the studied machine, no load: (*I*) PMs excited only without cans, (*II*) both PMs and armature coils excited without cans, (*III*) both PMs and coils are excited when cans are inserted ... 148

Fig. 5.16 Radial airgap flux density distribution, (**a**) without cans, (**b**) with cans ... 149
Fig. 5.17 Airgap flux density harmonics of the canned machine at different operation condition, (*I*) no load, (*II*) rated, (*III*) start ... 150
Fig. 5.18 The stator can loss variation as a function of the can radial length and material at typical operation conditions, (**a**) FEM, (**b**) analytical. In either case, *H* hastelloy, *S* stainless steel ... 151
Fig. 5.19 The rotor can loss variation as a function of radial can thickness value and material at typical operation conditions, (**a**) FEM, (**b**) analytical. In either case, *H* hastelloy, *S* stainless steel ... 151
Fig. 5.20 The can loss variation with rotor speed ... 152
Fig. 5.21 The can loss variation with rotor speed ... 153
Fig. 5.22 The rotor loss variation with rotor speed ... 153
Fig. 5.23 The torque-speed relationship with/without cans ... 154
Fig. 5.24 The torque and loss values under different durations of phase defluxing ... 154

Fig. 6.1 Introducing compensation elements t_{comp} into the traditional lumped parameter thermal network, FE is used as reference, (**a**) Temperature distribution by FE analysis, (**b**) The lumped parameter network model with compensation elements, (**c**) Numerical waveforms and comparison ... 158
Fig. 6.2 Geometrical overview of the studied canned machine ... 158
Fig. 6.3 Discretizing the machine into connected components ... 160
Fig. 6.4 A calculation example of thermal resistance ... 161
Fig. 6.5 Connecting resistances between stator teeth and yoke ... 162
Fig. 6.6 The contact coefficient relationship between the outer can and stator teeth ... 164
Fig. 6.7 Compensation elements by a couple of connected components with a contact coefficient ... 166
Fig. 6.8 Model of a coil, (**a**) Configuration, (**b**) Cut of a turn, (**c**) Heat flow ... 168
Fig. 6.9 Model of windings with X-Y-Z directions defined ... 169
Fig. 6.10 Heat flow in the machine end part ... 173
Fig. 6.11 Detailed heat flow of the proposed network model ... 176
Fig. 6.12 The network overview ... 177
Fig. 6.13 Temperature rise of each component due to the use of cans ... 178
Fig. 6.14 Detailed heat flow of the can shields and airgap ... 178

Fig. 6.15 Stator prototype of the canned machine, (**a**) canned, (**b**) before the stator can is fixed onto the bore, showing locates of thermal sensors, *1*: axially mid, *2*: axially partial, *3*: axially end, *4*: end windings. Note that plastic partitions that help to consolidate armature coils is removed before the stator can is fixed and thermal sensors will be attached on 1–3 of the can 179
Fig. 6.16 Nodal temperature rise of critical points, (**a**) under phase current $I_{ph} = 8.1$ A/mm^2, (**b**) under $I_{ph} = 16.1$ A/mm^2 180
Fig. 6.17 The calculation flow of the coupled analysis 181
Fig. 6.18 Can loss (**a**) and temperature rise (**b**) with rotation speed 0–2.5 krpm at steady state, the fixed turn-on/off control, with 170A hysteresis current level is applied 183

Fig. A1.1 Model of MMF distribution 190
Fig. A2.1 The phase current and harmonics 192

List of Tables

Table 2.1 Variation of width of Path 1 w_{p1} (mm) according to distance of pole tips AD (mm); The rotor position θ of 3-phase 12/8 SRMs is shown as reference, where 22.5° is unaligned position and 17.5° is where poles begins to overlap, with typical airgap lengths δ, SRM_I = 0.5 mm, SRM_II = 0.8 mm 17

Table 2.2 Variation of width of Path 1 w_{p1} (mm) based on the distance of pole tips AD (mm); The rotor position θ (Deg) of 3-phase 12/8 SRMs is shown as reference, where 22.5° is unaligned position and 17.5° is where poles begins to overlap, with typical airgap lengths δ, SRM_I = 0.5 mm, SRM_II = 0.8 mm 22

Table 2.3 The machine geometry variation based on M0 25

Table 2.4 The studied 3-phase 12/8 canned SRM and geometric parameter overview 27

Table 2.5 Comparison of different methods 34

Table 2.6 SRM volumetric constraints and features 39

Table 2.7 Local flux density (T) at saturated regions under phase current 200A 39

Table 2.8 Geometry overview of the studied SRM 43

Table 2.9 Geometric variation (in %) from the studied SRM in Table 2.8 as M0 48

Table 2.10 The machine geometry variation 52

Table 3.1 Comparison of eddy current loss on stator and rotor core of a standard and a canned SRM, S standard, C Canned 67

Table 3.2 Friction loss by coast down test, S standard, C canned 68

Table 3.3 Copper loss due to the use of cans, S standard, C canned 68

Table 5.1 The canned machine overview 141

Table 5.2 The can loss (W) from harmonics 144

Table 5.3 Loss (W) by different magnetic materials at no load 146

Table 5.4 The studied 12–10 canned SPM machine parameter overview 147
Table 5.5 The can loss generation (W) from harmonics by the proposed analytical model 150
Table 5.6 The simulated canned switched reluctance machine 152

Table 6.1 List of thermal resistances 159
Table 6.2 List of compensation elements 159
Table 6.3 List of heat sources 160
Table 6.4 List of thermal conductivities 160
Table 6.5 Thermal resistances in the machine end part 173
Table 6.6 List of thermal conductivities 174
Table 6.7 List of heat sources (time averaged values) 174
Table 6.8 Calculation results of all resistances in the thermal network 174
Table 6.9 Calculation results of all compensation elements in the thermal network 175
Table 6.10 Component temperature rise by different methods at steady state, with ambient temperature 23.4 °C 178
Table 6.11 Thermal dependent parameters of the can material 180
Table 6.12 Steady state can loss (both cans) and temperature variation of the stator can (on average) at different operation condition, with constant speed of 3 krpm 182

Chapter 1
Overview of Canned Electrical Machines

1.1 Background

A canned electrical machine is the core drive component in a hydraulic pump system, which has applications in nuclear power, coal or deep sea mining, as well as hydraulic energy conversion, etc. The pump body is connected with an electrical machine to form a drive system. The traditional structure is shown in Fig. 1.1a: The blade is driven by an ordinary machine and torque is delivered via a connecting mechanism that however leads to low efficiency, less reliability and complicated maintenance. Alternatively the improvement is shown in Fig. 1.1b that the machine is canned by adding a can-shield structure in airgap. The structure includes a stator can that is fixed onto the inner bore of the stator part and a rotor can onto the outer bore of the rotor part. The liquid being pumped is able to get into the airgap between cans. The liquid stays in the chamber walled by both cans and the axial end cap, and however cannot enter into rotor and stator slots. Configuration of the can shields is shown in Figs. 1.1c and d. Cans are resistant of high temperature, high pressure, erosion, and are zero leakage, which ensures durable operation for special liquid delivery under harsh environment.

As an application example [1], Fig. 1.2 shows the system integration with a canned machine. The liquid being pumped may get into the chamber inside the machine and is discharged out via the rear auxiliary pump mechanism, serving as coolant for can shields. Zero leakage, simple drainage and cleaning, duration, robustness and being reliable are expected features.

For contemporary electrical drives, low loss generation, high power density and reliability are fundamental requirements. In terms of a canned machine drive, low loss ensures high energy conversion from electrical to potential and momentum of the liquid being pumped. High power density enables efficient and compact mechanical structure. High reliability is of fundamental significance for safe operation and free from manual maintenance under harsh working environment. However when a

Q. Yu et al., *Analysis and Mathematical Models of Canned Electrical Machine Drives*, https://doi.org/10.1007/978-981-13-2745-2_1

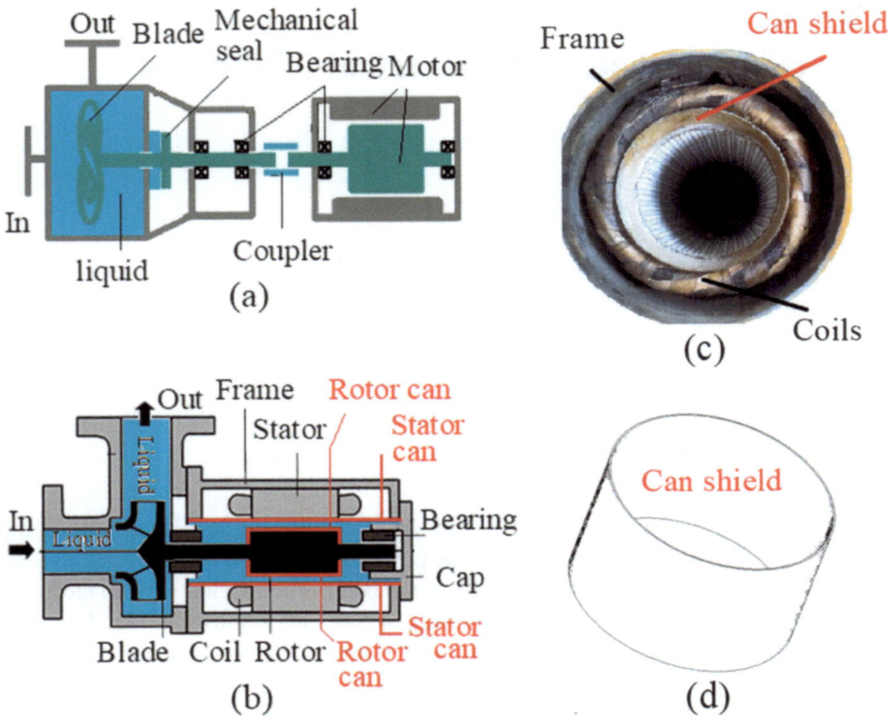

Fig. 1.1 Comparison of the traditional and improved hydraulic pump systems

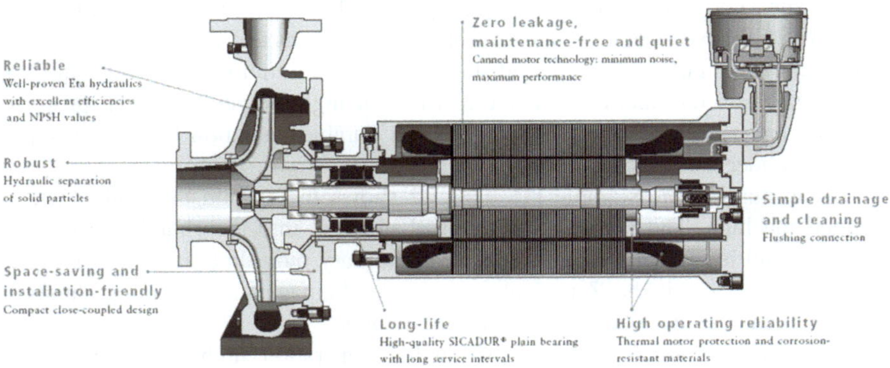

Fig. 1.2 The integrated system and the canned drive machine

machine is canned, all aspects are facing prominent challenges and details are as follows:

1. Loss generation

Due that the liquid being pumped exhibits potentially high temperature, high pressure, being poisonous, radioactive, erosive or scarcity, the can shields are manufactured using metal alloy. As cans are imposed in alternative airgap flux field, eddy current is induced, which further generates ohmic loss, here called CAN LOSS. Due to the manufacturing constraints, cans are produced a sleeve-like cylinder with very small thickness value, instead of the traditional stack lamination as rotor or stator. Therefore, the can itself is weak in resisting eddy current generation. It is found that [2, 3] consequent can loss is drastically higher than traditional copper or iron loss, leading to higher operation temperature and lower efficiency of the machine drive.

2. Power density

There are a couple of limitations for power density enhancement. For one hand, the use of cans increases radial thickness of airgap, and magnetic reluctance is increased that reduces flux density. For another, with increase of power density, the can loss rises up sharply, posing a key challenge on performance enhancement.

3. Reliability

The reliability refers to the degree of abrasion and aging of the can component, as well as system robustness, etc. The use of cans will to a large extent affect system reliability and reasons are as follows. For one hand, the can loss enhances working temperature in an escalated degree, which accelerates the aging of armature coils and poses risks such as rotor bar crack or demagnetization of permanent magnets, the temperature sensitive components when applied. For another, cans are source of high loss that contacts fast flowing fluid and there are great challenges on mechanical intensity, abrasion and aging.

In general, the use of cans leads to considerable ohmic loss, which further suppresses power density enhancement and reduces system reliability. Efficient management of can shields usage is the only approach for high performance of canned machines.

1.2 Research Overview

Due to characteristic features regarding can shield structure, electromagnetics and heat transfer of a canned electrical machine, literatures are not sufficiently available. According to expectations of low loss, high power density and reliability, at present, design and control principles are not systematically studied. Further, power density enhancement and reliability are not reported yet. Generally, the canned machine is partly developed in terms of structural topology, loss analysis, thermal and acoustic issues, and details are as follows.

1.2.1 Structural Topology

Due to working environment, it is inconvenient to apply armature coils with external controllable circuit in rotor part. Squirrel cage induction machine was first developed, followed by permanent magnet (PM) machine and switched reluctance machine (SRM). Typical studies are summarized as follows.

1. Induction machine

A squirrel cage induction machine of several kilowatts was canned as a hydraulic pump drive, which set up the way of developing canned motors [4]. Further, a 36-slot topology is reported [5] which reduces mechanical and electromagnetic interference to can shields by suppressing output torque harmonics and airgap magnetic field harmonics. In [6], the number of stator slots is further increased to 48 and meanwhile a double layer armature coil structure is used to reduce can loss by enabling a more sinusoidal airgap magnetic field. This canned motor is in megawatt as a main drive for nuclear power. As a classic prototype, squirrel cage induction machine is of simple structure, has wide application basis and is especially suitable for high power transfer. However, the inherent induction cage structure may take the risk of broken bars at high working temperature due to the can loss. Therefore, it is of great challenge to realize high power density and reliability features.

Due to its wide application, new topologies of induction machine are recently reported [7]. A phase-shifting design is studied, in which a couple of identical canned machines are equipped in series, with a can shield passing through the entire axial direction as a means of connection. The feature is that one of the machines is connected with a shift of circumferential angle by one stator slot. In analogy to lamination that suppresses iron loss by making sectors, the can loss by this method is reported a reduction by 23% with eddy current on the can shield confined in each sector. Due to the connection across both machines by cans, this topology is greatly challenged in terms of mechanical intensity. Also application of a couple of armature coils in this system is complicated that requires precise and coordinative control.

2. Permanent magnet machine

The permanent magnet (PM) brushless DC or synchronous machines adopt PMs in rotor that replace rotor armature coils, which is recently attended as a potential for canned machines. In [8, 9], a 0.7 kW small sized brushless DC canned machine is designed. This machine adopts DC excitation, has a higher starting torque and flexible control, and however consequent higher can loss is introduced as well. Alternatively in [10], a PM synchronous machine of an integral rotor structure is proposed. This machine generates reluctance torque by double saliencies of both rotor and stator. However, alternation of airgap reluctance may lead to prominent can loss generation. Such machine structure is suitable for small sized, instead of high power transfer. As an alternative in [11, 12], a widely adopted fraction slot concentrated windings (FSCW) 12–10 topology of a synchronous PM machine is canned. The machine applies surface mounted PM topology with increased output power up to 20 kW. Such topology helps to reduce can loss generation by suppressing airgap magnetic flux harmonics. However, the use of PMs has inherent shortcomings of demagnetization and weak robustness, as working temperature of a canned machine is in comparison drastically higher than a corresponding ordinary one. At the same time, the rotor contacts with shaft and airgap and therefore has worse heat dissipation ability, which further raises temperature on PMs. In the mentioned literature [8, 9], cooling liquid channels are designed in the rotor part, making complicated mechanical structure. Synchronous PM machine in the mentioned literature [11, 12] doesn't require such cooling, because the airgap length in radial direction researches high up to 7.5 mm to incorporate PMs, cans, and cooling channel, which to a large extent reduces airgap flux density.

3. Switched reluctance machine

Switched reluctance machines (SRMs) generate torque by magnetic reluctance variation, which has no magnetic excitation mechanism in rotor and thus is of high robustness. In [13, 14], a 3 kW canned SRM with low rotation speed is designed and tested. The problem to be solved is, due to DC excitation as well as vast fluctuation of airgap magnetic reluctance, harmonics is considerable and even more prominent than the working harmonic. The can loss generation is comparatively much higher than other type of machines of the same power range, leading to principal limitation on capacity and power density enhancement.

1.2.2 Loss Analysis

The can loss analysis serves as the fundamental part for design and optimization of a canned machine. Reported studies include empirical estimation, finite element (FE) method and analytical models and details are as follows.

1. Empirical estimation method

The can loss is estimated as a function of current excitation and rotor speed [4, 15]. This method uses an arithmetic expression that is able to roughly calculate the average value in a fast and simple way. However, it is subjected to feasibility only to a specific machine and often leads to inaccuracy when operation condition is subjected to complexity such as temperature jump or high phase current level. In [16], a magnetic equivalent circuit (MEC) model is developed for can loss calculation of an induction machine, which facilitates analysis of loss dependent parameters such as rotor slip and current excitation level. However, parameters in this model are

FE dependent which requires high computation. All these methods cannot model eddy current distribution on cans, one of the most important features when the machine is working.

2. Finite element method

The finite element (FE) method instead facilitates analysis of eddy current distribution on cans. In [17, 18], 2D and 3D FE analyses are respectively applied on a canned induction machine as a compressor drive, which illustrates that eddy current on cans at the axial end part is prominent. Further in [19], 3D FE is applied on a large sized canned induction machine as a nuclear power drive, in which the mechanical structure at the axial end part is designed by analyzing eddy current distribution on cans. In [20], the eddy current distribution by both stator armature coils and rotor excitations are analyzed. Although the FE method is able to model field spatial distribution, it cannot illustrate the generation mechanism of eddy current and loss.

3. Analytical methods

Analytical methods for canned machines are less reported. In [11, 12], magnetic field is analyzed starting from solving Maxwell equations. Magnetic vector potential distributed inside the machine is obtained, which is further used for flux and loss analysis. The feature is that the can shield cylinder is circumferentially divided into stripe sections. For each section, eddy current and loss are analyzed via Faraday Law of Induction. However, only eddy current excited by airgap flux field is considered whereas their electromagnetic coupling effect, a characteristic feature which will be studied later, is not fully described. In addition, this division of cans ignores rotor position dependent airgap reluctance variation and hence this model is only applicable for non-salient machine structures.

1.2.3 Thermal and Acoustic Analysis

In terms of vibration analysis, eddy current on cans induced by rotating magnetic airgap field will interact with that field to generate high-frequency electromagnetic excitation force. As the stator can sleeve is a thin-wall component and a tiny gap to the stator pole cap will lead to vibration and acoustic noise. With increasing demand for capacity, duration and reliability improvement, design of a canned machine is challenging, and mechanical analysis is necessary. In [21], modal analysis and mechanical deformation of can shields of a 25 kW canned induction machine is analyzed. In [22], an overall acoustic and noise reduction method of a canned machine is studied, including manufacturing inspection, design adjustment via key component assembly, excitation force and modal analysis. At the moment, an acoustic reduced machine design approach or control optimization scheme is not reported. In terms of thermal analysis, it is reported in [23] that variation of electrical resistivity and thickness of cans has great influence on temperature rise, and heat convection via stator frame has very limited effect on cooling. In [24], heat dissipation of a large sized induction machine for nuclear application is studied. Due to considerable can loss generation, temperature increases along the cooling channel. Further in [25], it is demonstrated the highest temperature locates at the machine axial end part.

1.2.4 Review Summary

As can shields are working in the sensitive airgap region, a canned machine has significant difference in terms of electromagnetic and mechanic performance. The cans are core component that affects airgap flux that leads to ohmic loss and further to thermal and acoustic issues. Recent studies show progress in terms of mechanical structure, loss and thermal analysis. However, there are still research fields that need further study as follows.

1. Novel mechanical structure

At present most literatures are all about squirrel cage induction machines. The induction machine has no PMs and less magnetic harmonics, making it suitable for large power application. However for medium size power transfer, throttling control is desired for higher efficiency that the machine is expected of frequent cranking and higher starting torque to drive liquid of viscosity and inertia. Under this assumption, the starting capability, robustness and speed regulation should be further improved and there are accordingly no novel mechanical structures reported.

2. Knowledge of the can shield dynamics

The study of using a can is about loss analysis and further thermal and acoustic. Conventionally, design of a canned machine simply falls into the update using a can shield on the basis of an ordinary one. The interference in terms of electromagnetic and thermal features due to use of cans is not fully studied, and hence characteristic features of a canned machine are not fully proposed.

1.3 Canned Switched Reluctance Machines

Rigid requirements have been made on canned machines. From the hydraulic load point of view, the regulation of fluid flow with a variable speed reduces loss and an inverter fed machine with variable frequency drive is desired. Classical solutions are induction machines connected with standard voltage supply and fixed frequency. The improvement was further made to drive with variable frequency. However, frequent regulation of speed leads to control complexity and low efficiency. Also in the frequent cranking, low rotating speed and high load application, adoption of induction machine is unsatisfactory or even incompetent.

There has been high demand in developing alternative topologies. The use of cans hardly allows energy supply in the rotor and therefore, rotor electrically excited machines are not considered. Switched reluctance machine (SRM) is a prospective solution as a pump drive, which can offer higher starting torque and wide constant power speed range [26–30]. The simple, robust rotor structure free from PM or rotor bars is suitable for high temperature environment and high-speed operation. SRM has an inherent degree of fault tolerance and high peak torque capability. Due to absence of PM and the use of tooth concentrated coils, the manufacturing cost is inherently low with well-established production procedures.

One characteristic feature of SRM structure is double saliency that both rotor and stator have prominent poles. Using this topology, each phase is in sequence fluxed and defluxed to generate reluctance torque. Accurate modeling of flux linkage curves and airgap flux distribution are the fundamental part for precise calculation of SRMs. However, the saliency structure leads to sharp change of equivalent airgap reluctance with rotor position, making it difficult to calculate. Further when power density is increased, severe magnetic saturation occurs. Such saturation concentrates at the pole tips, and is dynamically distributed according to rotor position and phase excitation level, which further complicates flux linkage and airgap flux field calculation. In this book, fast analytical models will be discussed.

When an SRM is canned, there exhibits some characteristic features in terms of electromagnetic and heat transfer aspects. In this book, electromagnetic and thermal features using a can will be discussed. Further, a couple of dynamic behaviors "electromagnetic coupling" and "electro-thermal coupling" of a can shield will be revealed. The electromagnetic coupling refers to the interaction between airgap flux field and the induced eddy current, the fundamental feature which determines the characteristic use of cans. For one hand, alternative airgap flux induces eddy current on cans and further generates ohmic loss. For the other, this current produces another magnetic field that overlaps with the original airgap field, which together affects electromagnetic performance. Such couple phenomenon is characteristic for canned machines and the key consideration in design, analysis and optimization.

Further on this basis, the electro-thermal coupling refers to the correlation between airgap flux and heat generation cans, another feature which determines characteristic use of cans. For one hand, the induced eddy current by alternating airgap magnetic flux determines loss and temperature rise of cans. For the other, electrical resistivity and heat conductivity of cans vary with the change of temperature, by which eddy current and further the airgap flux is in turn affected.

1.4 Outline Summary

In this book, electromagnetic features of an electrical machine are modeled and analyzed, with particular attention on saliency pole structure. Further when the machine is canned, the corresponding electromagnetic and thermal performance due to the can effect is discussed. Chapter details are as follows.

In Chap. 2, the electromagnetic saliency effect is modeled by means of making improvements on traditional magnetic equivalent circuits (MEC). Fast and accurate calculation of flux linkage is achieved on SRMs. Specifically, MEC is improved by discretizing the sensitive airgap region and high locally saturated poles into limited number of reluctance elements according to flux path characteristics. Such discretization accounts for both magnetic saturation and variation of airgap paths with small computation required. Alternatively, the discretization is replaced by using curve or surface fitting method to reduce complexity, with magnetic phase coupling considered. Further, the use of can shields is easily updated into the network model.

The canned switched reluctance machine makes a promising solution for hydraulic pump drives. Due to characteristic use of cans, operation principles and performances make a difference. In particular ohmic loss from cans is the principal consideration. In Chap. 3, the electromagnetic features and loss due to the use of cans are analyzed based on a salient SRM. First, at typical rotor positions, eddy current distribution on cans by a single phase, followed by all phases is analyzed. Next, consequent ohmic loss on cans is studied, that is extended to one stroke period and further to different control modes. Finally compared to a standard SRM, the feature of airgap flux and eddy current loss due to the use of cans is shown.

In Chap. 4, spatial distribution of magneto motive force (MMF) is studied, the prerequisite of analytical modeling of the can effect and canned machines. Armature coils as magnetic field excitation source are mainly studied, including both tooth concentrated and slot distributed topologies, followed by permanent magnets (PMs). In particular for concentrated coil, the wire arrangement inside is modeled. In addition, MMF in axial direction is studied.

In Chap. 5, an analytical model of concentric layer structure in 3 dimensions is proposed for canned electrical machines. This model divides the machine into airgap, cans, poles, yokes and shaft that are centered and connected cylindrical components. Based on MMF distribution in Chap. 4, this model uses Maxwell's Equations in the vector potential form. By separation of variables, magnetic vector potential of each layer is deduced. Loss of each layer including can loss is calculated by Poynting's theorem of energy conversion. Active torque of each layer is calculated by applying Maxwell stress tensor.

In Chap. 6, heat transfer and disssipation of a canned switched reluctance machine are analyzed. The model is based on traditional lumped network by adding compensation elements to reduce systematical mistake. In the past, such mistake is neglected or rounded by a curve fitting procedure. Such model requires small computation and is transferrable to electrical systems and electromagnetic devices. Further as an application, thermal-electro coupling effect on cans of a canned SRM is studied by combining the thermal and electromagnetic analytical models.

References

1. Pumps and Automation, *Product Catalogue* (KSB Aktiengesellsachaft, Frankenthal, Germany, 2009)
2. Annual report of KSB Aktiengesellschaft (Frankental, B.W., Germany, 2014), Available at: http://www.ksb.com/linkableblob/ksb-en/2589718-698722/data/Annual-report-2014-KSB-Group-data.pdf
3. Teikoku Co. databank (2015), Available at: www.tdb.co.jp/english/
4. R.C. Robinson, I. Rowe, Calculation of can losses in canned motors. Trans. Am. Inst. Electr. Eng. **76**(2), 312–315 (1957)
5. X.C. Zhang, W.L. Li, J.C. Cao, The can loss calculation and performance calculation performance analysis of a canned motor. J. Harbin Inst. Technol. **39**(9), 1422–1426 (2007). in Chinese
6. Y.P. Liang, Y.L. Hu, X. Liu, C.X. Li, Calculation and analysis of can losses of canned induction motor. IEEE Trans. Ind. Electron. **61**(9), 531–4538 (2014)
7. Y.J. An, Q. Zhang, W.R. Li, Y. Shi, Eddy losses and temperature rise on shielded can in canned motor with a new phase-shifted structure. J. Xi'an Jiaotong Univ. **6**(3), 401–406 (2014). in Chinese
8. D. Uneyama, Y. Akiyama, The proposal of can loss estimation method of a canned motor, International Conference on Electrical Machines and Systems, ICEMS 2007, Seoul, S. Korea
9. Y. Akiyama, Y. Niwa, K. Kawanisi, D. Uneyama, S. Manome, FEM analysis of relationships between gap Length and CAN's thickness and materials and loss, International Conference on Electrical Machines and Systems, ICEMS 2011, Beijing, China
10. Y.Y. Ni, Y. Huang, Q. He, Design and optimization of an integral rotor structure permanent magnet canned motor. J. Elect. Mach. Control **19**(12), 68–74 (2015). in Chinese
11. S. Urschel, *Entwurf und Optimierung energieeffizienter Pumpenantriebe auf der Basis von permanent-magneterregten Synchronmotoren in Zahnspulentechnik, Kaiserslauterer Beitraege zur Antriebstechnik, Band 2* (Shaker Verlag, Aachen, 2015). (in German)
12. Y. Burkhardt, G. Huth, S. Urschel, Eddy current losses in permanent magnet canned motors, International Conference on Electrical Machines, ICEM 2010, Roma, Italy
13. H. Yamai, M. Kaneda, Optimal switched reluctance motor drive for hydraulic pump unit, IEEE International Conference on Industrial Applications, IAS 2010
14. B.C. Kim, Performance analysis of a switched reluctance machine for hydraulic pump, International Conference on Electrical Machines and Systems, ICEMS 2005, Nanjing, China, pp. 659–663
15. S. Urschel, G. Huth, *Betriebsverhalten von Permanenterregten Spaltrohr Pumpen motoren* (VDI/ ETG-Fachtagung, Boeblingen, 2015), pp. 577–591 (in German)
16. L.T. Ergene, S.J. Salon, Determining the equivalent circuit parameters of canned solid rotor induction motors. IEEE Trans. Magn. **41**(7), 2281–2286 (2005)
17. K. Yamazaki, Modeling and analysis of canned motors for hermetic compressors using combination of 2D and 3D finite element method, International Conference on Electric Machines and Drives, IEMD 1999, Seattle, USA
18. L.T. Ergene, S.J. Salon, One-slot ac steady-state model of a canned solid rotor induction motor. IEEE Trans. Magn. **40**(4), 1892–1896 (2004)
19. Y.P. Liang, X. Bian, H.H. Yu, C.X. Li, Finite-element evaluation and eddy-current loss decrease in stator end metallic parts of a large double-canned induction motor. IEEE Trans. Ind. Electron. **62**(11), 6779–6786 (2015)
20. Y.Y. Ni, Y. Huang, L. Zhao, Three dimensional electromagnetic field and inductance calculation of a permanent magnet canned machine. Trans. China Electro-Tech. Soc. **30**(1), 98–104 (2015). in Chinese
21. Y.L. Yu, G.S. Zhang, D. Wu, The excited force and forced vibration analysis of the stator can for low noise canned motor. Trans. China Electro-Tech. Soc. **32**(S1), 160–169 (2017). in Chinese

22. C.X. Li, Technical research on the vibration problem for a new type of canned motor. Larg Electr. Mach. Hydraul. Turbine **1**(1), 18–21 (2015). in Chinese
23. C.X. Li, Z.D. Ren, Y.P. Lu, X.X. Kong, Influence factors research on three dimensional temperature field of the canned motor. Electr. Mach. Control **19**(8), 15–21 (2015). in Chinese
24. S.Y. Ding, B.C. Guo, H.J. Feng, Y. Zhang, H.T. Wang, C.G. Guo, Temperature field investigation of canned primary pump motors in nuclear power stations. Proc. CSEE **34**(9), 1368–1375 (2014). in Chinese
25. W.J. Chen, L.J. Miao, S.Y. Ding, Thermal calculation of the stator end part of a can- shielded motor. Electr. Mach. Control **20**(2), 83–97 (2016). in Chinese
26. A. Emadi, L.J. Young, K. Rajashekara, Power electronics and motor drives in electric, hybrid electric and plug-in hybrid electric vehicles. IEEE Trans. Ind. Electron. **55**(6), 2237–2245 (2008)
27. B. Bilgin, A. Emadi, M. Krishnamurthy, Design consideration of switched reluctance machines with higher number of rotor poles. IEEE Trans. Ind. Electron. **59**(10), 3745–3756 (2012)
28. A. Chiba, Y. Takano, M. Takeno, Torque density and efficiency improvements of a switched reluctance motor without rare-earth material for hybrid vehicles. IEEE Trans. Ind. Appl. **47**(3), 1240–1246 (2011)
29. H.B. Ertan, *Modern Electrical Drives* (Springer, Berlin, 2000), pp. 141–195 ISBN:978-94-015-9387-8_8
30. M. Ahmad, *High Performance AC Drives* (Springer, London, 2012), pp. 129–160 ISBN:978-3-642-13150-9_6

Chapter 2
Electromagnetic Analysis of Saliency and Can Effect by Network Models

2.1 Flux Linkage Modeling of Switched Reluctance Machines

Determination of flux linkage characteristics is a central step in electromagnetic modeling. The flux linkage, a group of curves varying with phase current levels and rotor positions, is required very accurate, on which main outputs such as torque, radial force and iron loss highly depend [1, 2]. However, modeling of flux linkage, a seemingly simple question, is still under study, especially for electrical machines with salient poles. This is because of the saliency that leads to rotor position dependent airgap paths together with magnetic saturation. Further in this study, flux linkage curves are complicated by adding can-shields in airgap. Due to electrical resistance of cans, the alternating airgap flux will induce eddy current circulation which will in turn affect that airgap flux, making a coupled relation.

Switched reluctance machine (SRM) is an attractive electrical propulsion drive with simple and robust structure, low cost and high speed capability. One of the trends is to increase power density by using higher phase current excitation level, causing severe magnetic saturation [3]. Compared with other machines, magnetic saturation of SRM is more obvious when the same power density is reached. Due to double saliency structure, the magnetic saturation layout is quite unbalanced. The saturation distribution is regarded varying with both phase current excitation level and rotor position, making magnetic flux curves quite nonlinear. In most cases, magnetic saturation occurs when an exciting pole pair begins to have alignment, as there exists the lowest airgap reluctance that facilitates flux lines passing through. For that case, tips of poles close to alignment are severely saturated if high current is imposed, and accurate estimation of magnetic curves at these rotor positions is critical.

Typical analytical approaches are classical deduction [4, 5] or curve fitting methods [6]. In [5], calculation of flux linkage of SRMs resorts to Ampere's Circuit Law and is separately modeled according to whether the pole pair has degree of

Q. Yu et al., *Analysis and Mathematical Models of Canned Electrical Machine Drives*, https://doi.org/10.1007/978-981-13-2745-2_2

alignment. In either case, magnetic saturation is included by an iterative progress. However if taking a close look at flux linkage curves especially when teeth are partially or above to overlap, there is still a small, but decisive discrepancy in value, which however would make severe deviation if further analysis is based upon. In [6], the curves were estimated using an exponent $\lambda(\theta, i) = a_1(1-e^{a2i}) + a^{3i}$, with coefficients a_1–a_3 that respectively represents flux linkage as the motor enters saturation, severity of saturation, and incremental inductance at high current. a_1–a_3 depend on position as well as geometry and were stored by Fourier methods. However, presence of higher order harmonics adds ripple to predicted instantaneous torque production.

Magnetic equivalent circuit (MEC) network model is a classic solution [7, 8] in which the machine is divided into connected components and each is represented by a lumped magnetic reluctance element. However, such model is less accurate on SRMs. Large slots on both rotor and stator make equivalent airgap vary with rotor position. Hence actual airgap reluctance does not necessarily equal to airgap length. Further, the simplification of using one reluctance element on each pole, as often applied on other electrical machines that do not have obvious saliency [9–12], is subjected to computational discrepancy. If such simplification is applied, flux curves obtained may coherently fit other means as exception cases at the aligned, unaligned rotor positions and their neighborhood [13, 14]. This is because the exciting poles are unlikely saturated at unaligned position while tends to get uniformly saturated throughout close to the aligned position. Magnetic resistance of flux paths due to saturation effect, a determination factor for flux linkage calculation, is often estimated by a lookup approach. The data is typically obtained via finite element (FE) method [15], and helps to achieve high calculation accuracy when being applied in models developed. Alternatively measurements are used [16, 17]. The lookup approach simplifies the estimation of nonlinearity. In [16] updates can be made conveniently to include iron loss into the network model proposed.

Besides flux linkage characteristics, it is necessary to investigate subsequent iron loss in a fast and convenient way, before making a prototype with a specific lamination material. This would help making a compromise between the cost and performance, depending on the application. In a magnetic material, iron losses occur when the material is subjected to a time varying magnetic flux density. Iron loss has two main components: hysteresis and eddy current loss. Hysteresis losses are the result of the rotation of the magnetic particles to line up with the external magnetic field. This produces molecular friction and, hence, energy consumption. When the traction motor is working at low speed, the hysteresis loss is the main part of iron loss, as the motor is exposed to high excitation current. Due to alternating flux density, eddy currents are induced in laminations and when combined with the limited conductivity of electrical steel, they cause resistive losses. When the machine is working at high speed, the eddy current loss takes up the main part of iron losses due to the high rate of change of flux density.

Different analytical methods have been proposed to predict iron losses. Generally they can be classified into loss separation model [18–20], the energy vector model [21, 22], and improvements from Steinmetz model [23–27]. In the separation model, iron losses are divided into the losses due to linear magnetization, rotational magnetization and higher order harmonics. However, the loss coefficients in this model usually depend on experiments. The energy vector model can provide a more detailed solution. The model starts by solving Maxwell Equations. With magnetic vector potential obtained, the flux density distribution in the air gap is determined, which paves the way for loss and torque calculation [28–30]. However, an infinite permeability of iron is considered for simplification when compared that of air. This is not accurate for SRM, especially at high current excitation level, when considering its double saliency structure. In Steinmetz model, the hysteresis and eddy current losses are separately calculated based on the variation in flux density. The correction factors [31] or excessive loss [32, 33] are introduced if current excitation is not sinusoidal. However, the Steinmetz method is less accurate when applied in SRM. Under high current excitation as in a traction application, the flux density and frequency varies greatly in different locations of the machine. As a result, the loss density is extremely unevenly distributed. Furthermore, as the flux density and frequency varies, and the loss coefficients are not constant anymore.

In this chapter, circuit network based models are proposed to analyze the magnetic saliency effect of SRMs, including flux linkage characteristics and iron loss. Organization is as follows. In Sect. 2.2, a novel magnetic equivalent circuit (MEC) model is developed to model the magnetic flux path with distribution of magnetic saturation considered. The feature is that, flux path is divided into branches based on airgap flux distribution. Each branch is further anayzed according to variation of magnetic saturation. Contributions are made on detailed modeling and determination of magnetic flux path dimensions whether or not poles have alignment, and on clarifications related to how to calculate and manage pole reluctance variation effectively. Further, this model is updated by adding cans to analyze flux linkage characteristics due to the can effect.

In Sect. 2.3, this model is further developed to estimate iron loss stemming from non-sinusoidal current excitation, based on definition of hysteresis and eddy current losses. By considering minor BH loops, the accuracy of hysteresis loss calculation is improved. The excessive loss, which is usually estimated using empirical equations, is not needed. Similar to finite element analysis, the machine is discretized, to account for unevenly distributed flux density, into very limited number of elements. Hence, the computational effort is greatly reduced. Variable loss coefficients are applied according to derived magnetic flux densities in different parts of the machine.

In Sect. 2.4, a FE assisted network model is proposed further to reduce calculation complexity of airgap path and in particular with phase magnetic coupling considered. The MEC is for the rotor position dependent airgap reluctance calculation while the FE part is for magnetic saturation estimation. Calculation accuracy in terms of flux linkage is discussed, followed by torque generation considering the magnetic coupling.

Alternatively in Sect. 2.5, accuracy improvement on airgap flux density modeling is described based on a novel magnetic circuit network. The calculation is featured by taking conductor layout and slotting effect. Emphases are made on airgap permeance and magnetic saturation identification using a numerical fitting method. The fitting of airgap resistance is enabled by a fourth order polynomial while magnetic saturation of saliency is fitted by a three dimensional surface.

2.2 A Discretized Circuit Network Model

2.2.1 Modeling of Airgap Reluctance

The proposed model is divided into a couple categories according to whether poles have alignment, as effective airgap path differs. Determination of airgap reluctance is decisive, as permeability of air is very small, even against highly saturated steel. The reluctance is calculated based on pole geometry and rotor position only, where no empirical or approximation equations are used. For either category, the following assumptions apply,

1. Each path is recognized by straight lines or arcs.
2. The shortest way principle: the flux chooses to go into rotor from stator poles through the minimum length, through flux leakage perpetuates.
3. The airgap curvature is considered.

2.2.1.1 Non-overlap Positions

This category applies from unaligned position to where poles begin to overlap. Figure 2.1 shows the flux path model. All airgap flux can be collectively represented by 7 paths where 1–3–4 and 5–6–7 are similarly arranged. As rotor moves counter-clockwise, Path 1 becomes the shortest distance whereas Paths 5–6–7 are gradually losing influences. In Fig. 2.1, Point O stands for central axis, and Straights $\overline{OA}$, $\overline{OC}$,

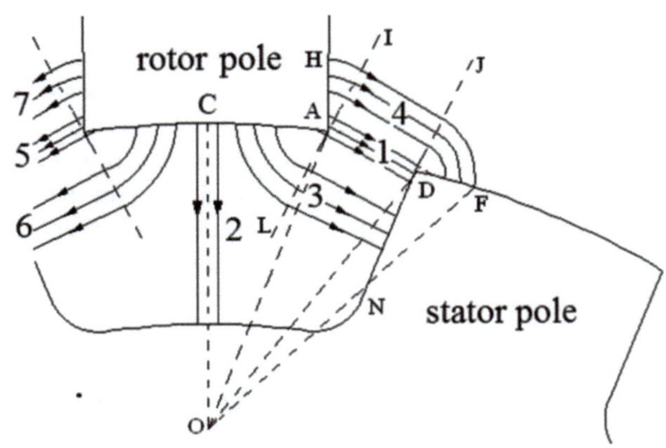

Fig. 2.1 Definition of airgap magnetic flux paths when poles don't have degree of alignment

OD and OF equal to outer radius of rotor poles. Points A, I, and L are on the same line that is vertical to flux Path 1. DJ is in parallel to IL.

Path 1, which is defined as the shortest way between pole tips AD, is the main path although not as decisive as when poles overlap. Geometrical determination of Path 1, including the length and width, serves as the start that other path relations are linked with. Length identification is simplified by averaging all the 3 flux lines of Path 1 nominally shown in Fig. 2.1, which is reduced to AD as well as the path width that will be illustrated later. Other Paths include 2–3–4. Path 2 represents slot leakage and 3–4 represent in-profile flux. As to length of each path, Path 3 is modeled by an arc bounded by AC and IL and then a connecting straight in parallel to Path 1. The length of 2–4 can be easily obtained. Then, the width of each path is modeled. The widths of Paths 2–3–4 are mutually adjusted according to the shortest way principle. Specifically, the boundary location between 2 and 3 is obtained by letting identical the lengths of their boundary connecting flux lines. In analogy, the boundary between 1 and 3, and between 1–4 is calculated. Hence, path widths of 2–3–4 are known. Paths 5–6–7 can be likewise calculated.

All airgap paths are now known except the width of Path 1 defined as w_{p1}. Identification of w_{p1} is carried out by a fitting method. Assuming that iron permeability is infinite, in such a linear case the magnetic curves as target data can be easily obtained. When a w_{p1} value is given, magnetic curves are calculated and compared with target data. To illustrate such fitting method, a couple of SRMs having different airgap lengths δ (0.5 mm and 0.8 mm) and pole widths is studied. The w_{p1} variation with rotor position θ is shown in Table 2.1. In either case, w_{p1} rises linearly with rotor from unaligned rotor position 22.5° till where poles begin to overlap 17.5°. The relation between w_{p1} and AD is further shown in Fig. 2.2. It is found that variation of

Table 2.1 Variation of width of Path 1 w_{p1} (mm) according to distance of pole tips AD (mm); The rotor position θ of 3-phase 12/8 SRMs is shown as reference, where 22.5° is unaligned position and 17.5° is where poles begins to overlap, with typical airgap lengths δ, SRM_I = 0.5 mm, SRM_II = 0.8 mm

	θ	22.5	21.5	20.5	19.5	18.5	17.5
SRM I	AD	3.593	3.07	2.561	1.912	1.294	0.848
	w_{p1}	0.920	1.591	2.212	3.081	3.883	4.602
SRM II	AD	3.777	3.122	2.557	2.029	1.488	1.003
	w_{p1}	0.854	1.599	2.252	2.899	3.651	4.446

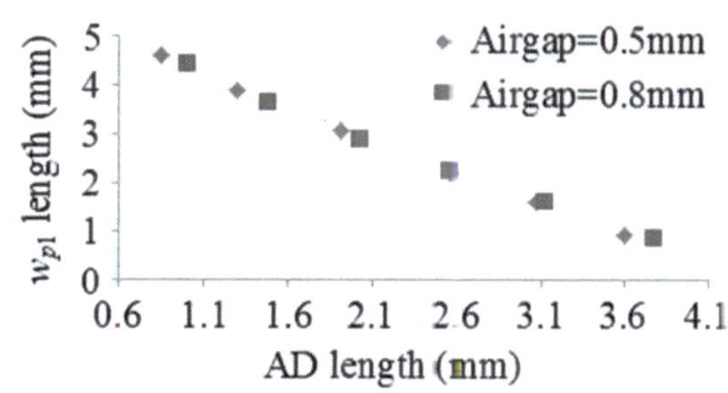

Fig. 2.2 The relation between physical distance of pole tips AD (mm) and modeled width of the main flux path w_{p1} (mm) when poles don't have alignment as depicted in Fig. 2.1, showing independence of w_{p1} to machine geometry variation

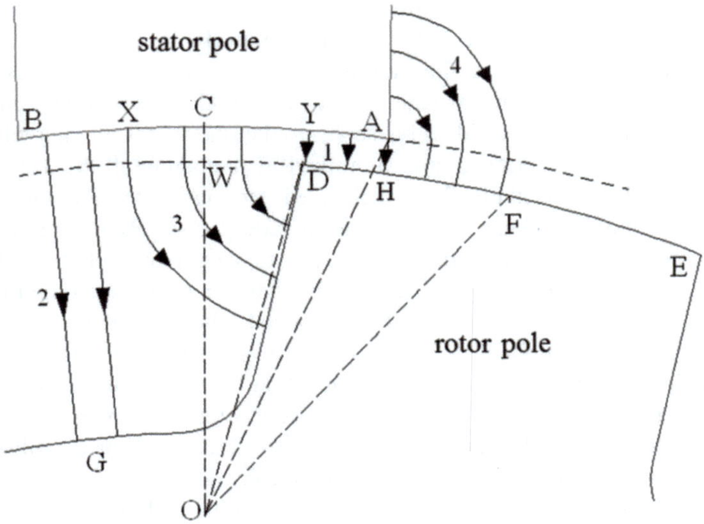

Fig. 2.3 Definition of airgap magnetic flux paths when poles have degree of alignment

all points is close to linear, and that is, relies on distance of pole tips $\underline{AD}$ regardless of pole shape and airgap length δ. Therefore, such fitting method can be used under various machine geometries.

2.2.1.2 Partial and Full Overlap Positions

This category range falls into where rotor pole starts to overlap till full alignment. Main flux lines are modeled in Fig. 2.3. There are together 4 featured paths and Path 1 is decisive, with width of overlap $\underline{AY}$ or $\underline{DH}$ and airgap length δ. Such case is simple, as Path 1 can be easily determined. In analogy, Path 2 stands for slot leakage flux while Paths 3–4 for in-profile flux in the form of straights in airgap while arcs in slots. The widths of Paths 2–3 are distinguished likewise by the shortest way principle. With degree of overlap increasing, the width value of Path 2 reduces till 0.

2.2.2 *Modeling of Pole Reluctance*

After modeling of airgap reluctances, reluctances on the machine body are illustrated. Yoke fluxes tend to uniformity and the yoke part is unlikely to get saturated, whereas pole fluxes are unbalanced and time varying and thus the pole part is subjected to high local saturation. Hence, modeling of pole reluctances is a key point. To account for flux density distribution characteristics, pole reluctances are

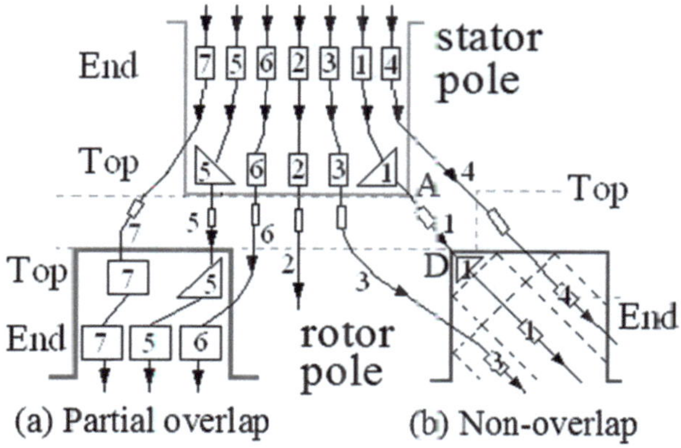

Fig. 2.4 Discretizing pole reluctances according to airgap paths, (**a**) partial overlap, (**b**) non-overlap

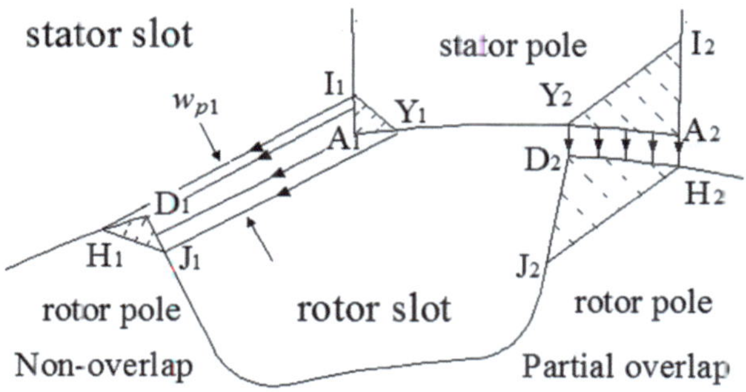

Fig. 2.5 Determination of pole tip reluctances that linked with the main airgap flux path at non-overlap and overlap rotor positions

discretized according to distribution of magnetic saturation. The arrangement when poles are respectively aligned and not aligned is shown in Fig. 2.4. It is stressed that each airgap reluctance block connected with a flux line represents one characteristic flux path as shown in Figs. 2.1 and 2.3. Hence, Fig. 2.4 shows the extended pole reluctance modeling based on airgap flux paths. To be specific, both rotor/stator poles are discretized into the top and end sections. When each flux line goes through, a top and end reluctance is attached. Note that for in-profile flux Paths 4–7, top reluctance is omitted. Also note that when poles are not aligned, Paths 3–4 in rotor pole are in parallel to Path 1.

For all top reluctances in Fig. 2.4, each size can be fixed once the corresponding airgap reluctance is known. For block reluctance 2–3–6, both the width and length are set identical to width of the known airgap reluctance on each flux path. Note that the pole tip reluctances 1–5 with an isosceles triangle are featured, indicating highly saturated zone of magnetic saturation. Dimension of the triangles can be found from a simple geometrical relation in Fig. 2.5. When poles are not aligned, the triangle bottom lines I_1Y_1 and H_1J_1 equal to the known widths of airgap flux w_{p1}. When poles have alignment, the isosceles side equals to the overlap width A_2Y_2 or D_2H_2.

For all end reluctances in Fig. 2.4, each size is first obtained and then the width values will be optimized by an iterative program. For stator pole at initialization

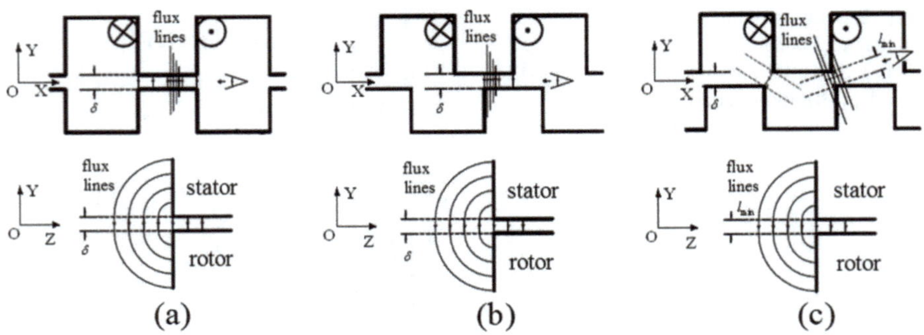

Fig. 2.6 Flux lines at the end part, (**a**) fully aligned, (**b**) partial aligned, (**c**) non-overlapping

stage, each width of end reluctance equals to the divided stator pole width by number of flux paths. Similar assumption can apply on rotor poles, however with modification when poles are not aligned, as flux line doesn't go vertically, expanding equivalent path width. As to length of each end reluctance element, the value is fixed by subtracting the length of corresponding top reluctance from the total pole height.

2.2.3 *Modeling of the End Part*

Flux leakage at the machine axial end part exhibits more obvious for canned machines. In analogy, such flux distribution is modeled by a simple method. The geometric relation at the representitive rotor positons are shown in Fig. 2.6. Paths are presented by arcs and straights in parallel with airgap lines. The path for each type is quite similar and the main difference lies in width and length of airgap reluctance according to rotor positions.

2.2.4 *Calculation Flow*

The effect of nonlinear magnetization characteristics of lamination material, in this model in terms of magnetic permeability variation for each reluctance element, is taken into account. Meanwhile, the distribution of magnetic saturation is considered by means of estimating equivalent magnetic reluctance by adjusting geometrical dimensions of pole tip and end reluctances. Both aspects are realized by an iterative algorithm of a dual-loop structure in Fig. 2.7, in which the machine geometry, rotor position and current excitation are input as prerequisite. Starting from an initial permeability value for all reluctance elements, the inner iteration loop is satisfied by convergence of permeability to account for magnetic severity. Further, the outer loop

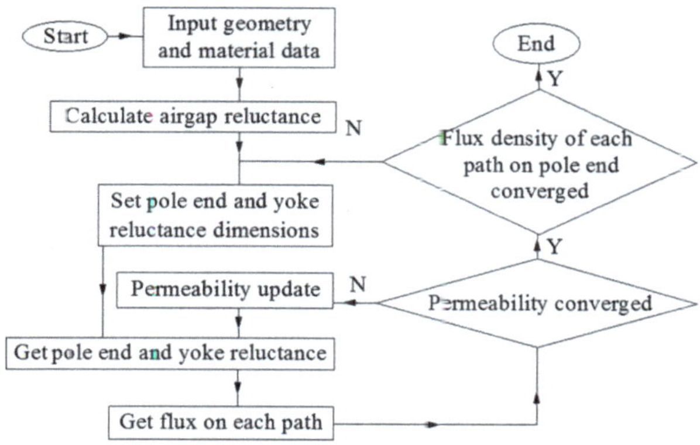

Fig. 2.7 The calculation flow with a dual-loop structure

is satisfied in terms of an identical flux density value throughout the end section of the pole by adjusting widths of the end reluctances.

The approach of discretizing and determining pole reluctance values is justifiable according to characteristics of magnetic flux. After airgap flux paths are determined, the size of pole reluctances on top section is simultaneously obtained as well, assuming that all flux lines travel the same direction as they are just before going into the pole. Then, flux lines subject to rearrangement of direction till parallel before going into yoke. When all lines have been rearranged, widths of paths at pole end are subjected to difference, which is modeled by adjusting width of the end reluctances

2.2.5 *Application Examples*

A 30 kW 4-phase 8/6 switched reluctance machine is considered. Based on the studied machine referred as M0 ($\beta_s = 20°$, $\beta_r = 20°$, $\delta = 0.5$ mm), M1 ($\beta_s = 30°$), M2 ($\delta = 0.25$) and M3 (both $\beta_s = \beta_r = 30°$) with specified modifications are applied. Principal variation includes airgap length δ, widths of rotor and stator poles β_s and β_r, to verify the calculation adaptability.

The calculation accuracy regarding airgap reluctance is verified by setting magnetic permeability of iron infinite. Figure 2.8 shows the airgap permeance versus rotor position by the proposed method with finite element (FE) analysis as reference, within the full rotor position range from the aligned 0° to unaligned 30°. At the time when the mentioned M1–M3 variation that is sensitive to airgap permeance takes effect, the proposed network to model airgap paths is seen adequate, as samples (dots) from a series of FE calculation are consistently close to the analytical curves. It is also shown that the inductance change due to pole width variation is obvious. In M1 when one pole width of a pole pair is increased, the rotor position where the inductance starts to drop is postponed. When both pole widths are enlarged (M3), the phase inductance close to 0° increases. When airgap radial length δ is reduced (M2), the inductance value is increased when poles have alignment.

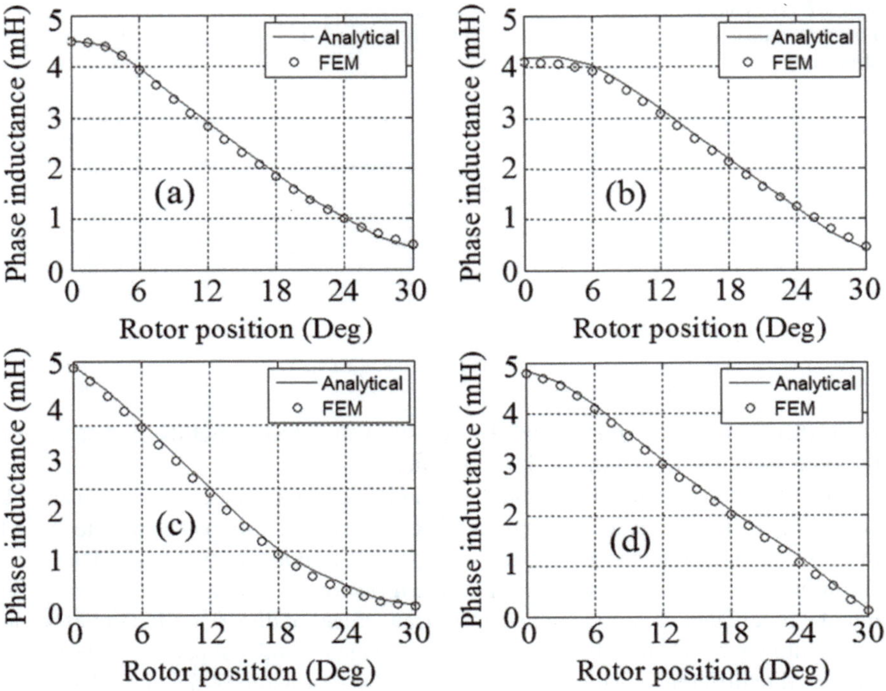

Fig. 2.8 The flux paths and airgap permeance variation of a pole pair by FE and the proposed analytical network. Note that in FE, electrical steel does not saturate. Main geometrical variation of the machine includes (**a**) the original machine, (**b**) with increased rotor pole width M1, (**c**) with enlarged airgap length M2, (**d**) with both increased stator and rotor pole widths M3

Table 2.2 Variation of width of Path 1 w_{p1} (mm) based on the distance of pole tips AD (mm); The rotor position θ (Deg) of 3-phase 12/8 SRMs is shown as reference, where 22.5° is unaligned position and 17.5° is where poles begins to overlap, with typical airgap lengths δ, SRM_I = 0.5 mm, SRM_II = 0.8 mm

	θ	22.5	21.5	20.5	19.5	18.5	17.5
SRM I	AD	3.593	3.07	2.561	1.912	1.294	0.848
	w_{p1}	0.920	1.591	2.212	3.081	3.883	4.602
SRM II	AD	3.777	3.122	2.557	2.029	1.488	1.003
	w_{p1}	0.854	1.599	2.252	2.899	3.651	4.446

Further to verify calculation of magnetic saturation, the electrical steel M1929 is taken. The number of turns per pole is 35 and phase current can reach up to 120A, making the magnetic saturation sufficiently high. Flux linkage curves are distinguished by an increment of 1° rotation, indicating high calculation accuracy is necessary. Typical machines M0, M4 and M5 are shown in Table 2.2, where M0 is the original machine. Variations of airgap length δ and pole widths β_r are respectively highlighted.

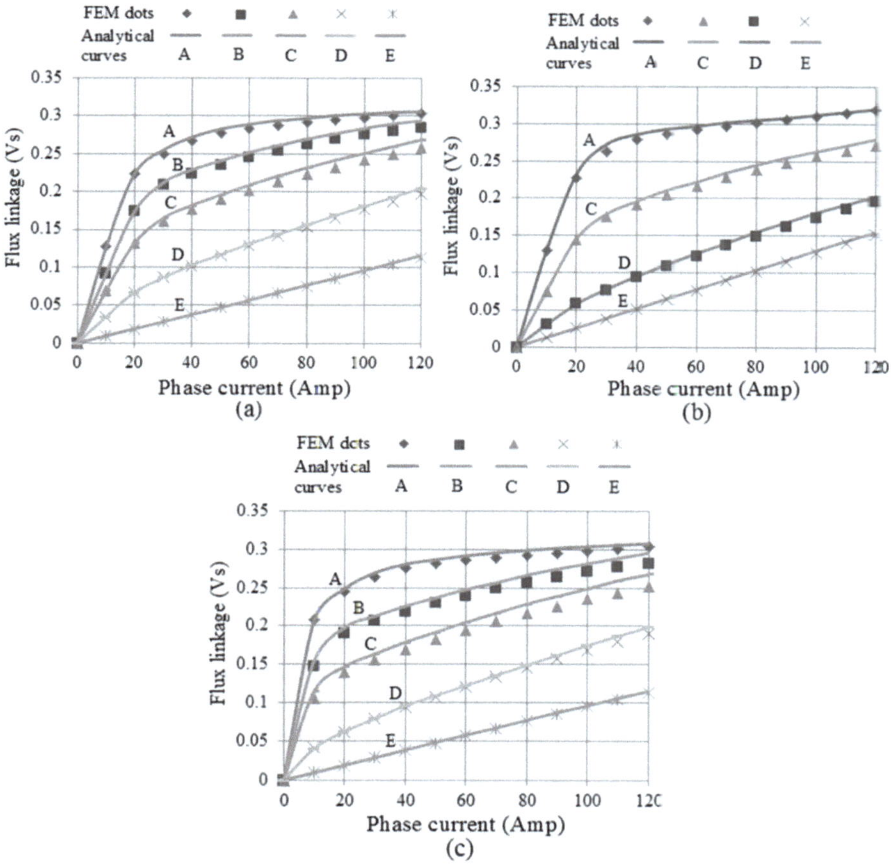

Fig. 2.9 Flux linkage characteristics of machines in Table 2.2, (**a**) original M0, in which A = aligned, B = partial overlap, C = just after overlap, D = just before overlap, E = unaligned, (**b**) increased rotor pole width M4, in which A = aligned, C = just after the start of overlap, D = just before the start of overlap, E = unaligned, (**c**) reduced airgap length M5, in which A = aligned, B = partial overlap, C = just after the start of overlap, D = just before the start of overlap, E = unaligned. Dots are FE data while curves are analytical data

Magnetic curves by both the proposed network model and FE are calculated, starting from aligned position 0° with an increment of rotor position 1° till unaligned 30°. Selected magnetic curves of typical rotor positions are comparatively shown in Fig. 2.9, including rotor positions of the full aligned, partial overlap, just after the start of overlap, just before the start of overlap and the totally unaligned. It is shown that each curve by the proposed network fits with FE calculation. At unaligned position where the magnetic saturation unlikely happens (Line E), both methods agree, indicating high calculation accuracy of airgap reluctances. At the aligned position (Line A), all flux paths goes through poles perpendicularly, which simplifies division of pole reluctance, as magnetic saturation is assumed uniformly distributed on poles. Note that there is slight numerical deviation when curves are bending, i.e. at 20–50A, showing calculation accuracy of magnetic permeability of pole reluctances. The accuracy can be improved by enhancing convergence condition, but at cost of computation time. As to curves at rotor position where poles are just before or after start of overlap (Line C and D), there exists a rising discrepancy with increase of phase current. High saturation may occur on tips of poles and hence, the limited number of pole reluctances proposed could not cover exactly high gradient variation of pole flux density, resulting in error when magnetic saturation is prominent. Such discrepancy can be reduced by using more characteristic pole tip reluctance whose geometry varies with rotor position and phase current, but at cost of computation time as well. In most cases, the proposed pole discretization approach is sufficient for a wide power range. When pole are partially aligned (Line B), such discrepancy reduces, as the main flux path that goes vertically through airgap and consequent pole reluctances especially on pole tips are easier to estimate, even under high current.

Phase torque is further analyzed to verify calculation accuracy. Torque is based on flux linkage curves that are differentiated by 1° rotor position, and thus likewise high accuracy is required. Torque waveforms with current variation are shown in Fig. 2.10 according to geometrical variation in Table 2.3. Both the proposed network model and FE agree at low phase current level while some discrepancy occurs with current increasing, but less than 10%. The most deviation occurs when the machine is highly saturated, due to the limited number of discretized elements.

Accurate modeling of magnetic flux characteristics of switched reluctance machines with high power density is challenging due to the double saliency structure, leading to variable airgap paths and distributed magnetic saturation. The proposed method divides airgap paths according to rotor positions and estimates distribution of magnetic saturation on poles with limited number of elements. Fundamental magnetic features can be estimated without empirical equations or FE method. It can calculate the flux linkage characteristics efficiently, especially at intermediate rotor positions. The method can be applied to various SRM geometries without power or speed limit.

2.2.6 Can Loss Analysis

The proposed network circuit model is updated with a stator and rotor can shield by simply setting electrical resistance of the can material. Figure 2.11 shows the network sketch at a typical patrial aligned rotor position. According to Faraday Law of Induction, airgap flux will lead to eddy current induction on cans that serves as additional source of magneto motive force (MMF) in the circuit. That is, the alternative flux that is deemed vertically penetrates through airgap will generate eddy current on the can of a certain thickness. Hence, it is predicted that flux values when cans are inserted will reduce and in such a way the cans serve as a source of reversal MMF.

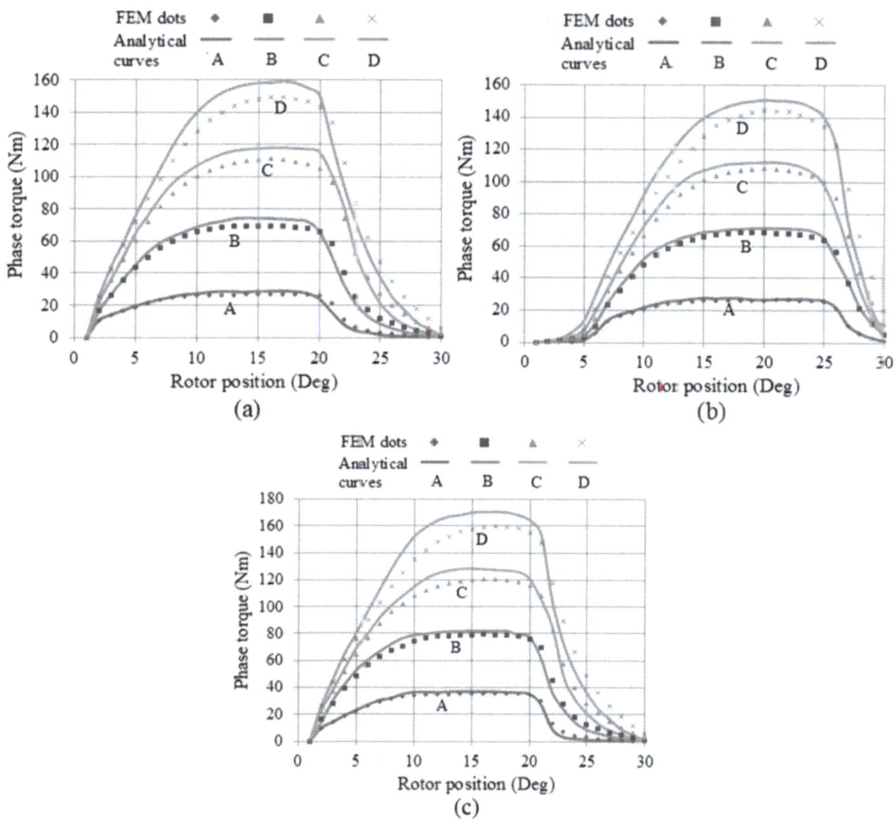

Fig. 2.10 Phase torque curves of machines in Table 2.2, with phase current variation, (**a**) original M0, (**b**) increased rotor pole width M1, (**c**) reduced airgap length M2. In all cases, phase current levels (in ampere) are, A = 30, B = 60, C = 90 and D = 120. Dots are FE data while curves are analytical data

Table 2.3 The machine geometry variation based on M0

	N_s	N_r	β_s	β_r	δ
M0	8	6	20°	20°	0.5 mm
M4				30°	
M5					0.25 mm

To have a magnetic field overview, the machine is also modeled by FE method in 3 dimensions, in which cans are highlighted. In Fig. 2.12a. Due to extremely thin airgap that makes flux variation sensitive, cans are finely meshed. That is, each can cylinder is circumferentially decomposited into up to 360 pieces as boundaries to limit mesh sizes. Therefore, most elements to be sovled locate in airgap region. The double saliency makes complicated airgap flux path, together with high local magnetic saturation. Further Fig. 2.12b shows an energized pole pair of partial alignment with emphasis on both cans. The flux paths, including the main and

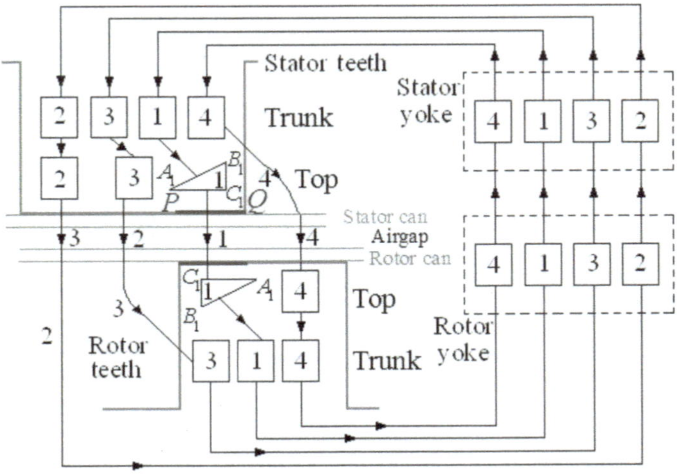

Fig. 2.11 The network update by adding rotor and stator cans, and note that the radial length values of airgap and cans are disproportionally enlarged

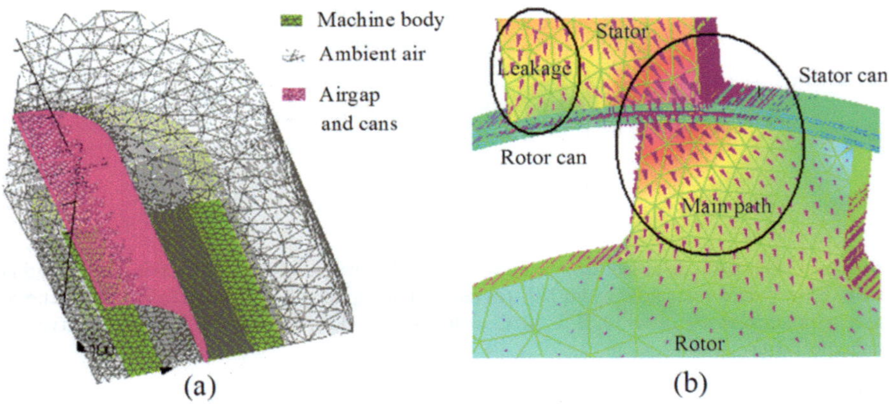

Fig. 2.12 FE modeling and analysis of a canned switched reluctance machine, (**a**) Geometrical overview with can shields highlighted, (**b**) detailed flux paths and density distribution

leakage, go through each can shield, inducing eddy current swirls that lead to electromagnetic coupling in airgap. Therefore, flux linkage features will make a difference.

On the analytical network model, flux curves are realized by an iterative program in Fig. 2.13, in which the machine geometry, rotor position and phase current information are identified as prerequisite to analyze all the 1–4 paths shown in Fig. 2.11. Meanwhile, auxiliary paths at the machine end part shown in Fig. 2.6 are taken simultaneously. To precisely calculate all paths, a couple of sections, including magnetic reluctance identificaiton and equivalent reversal MMF

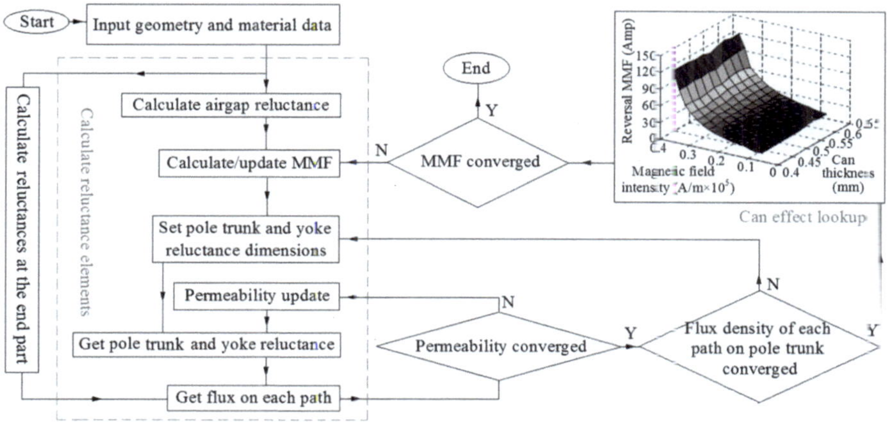

Fig. 2.13 The calculation flow

Table 2.4 The studied 3-phase 12/8 canned SRM and geometric parameter overview

The machine		The can and airgap	
Rated power (kW)	7.5	Stator can thickness (mm)	0.3
Base speed (rpm)	1500	Rotor can thickness (mm)	0.3
Max. speed (rpm)	4500	Airgap length (mm)	1.0
Bore radius (mm)	67.5	Stack length (mm)	61

estimation, is carried out. The first section refers to reluctance calculation of each path, including airgap, pole trunk and yoke, as well as pole top. It is shown that geometric widths of pole trunk and yoke reluctance elements are variable, which is numerically adjusted by a convergence condition based on the fact that flux density value of each path is identical. Meanwhile, reluctance values at pole top are iteratively obtained by magneitc permeance update. The second section refers to the can effect that the induced eddy current interacts with airgap flux, a complicated procedure that is deemed as an equivalent source of MMF that hinters airgap flux variation. Based on the law of magnetic induction, the reversal MMF value is taken as a function of magnetic intensity as well as thickness of the metal shield that airgap flux penerates through, which is estimated by a lookup table. The MMF is considered linear to the shield thickness and approximately quadratic to magnetic intensity variation. Due to similar imposure in airgap flux environment, both cans are merged as one component in the lookup.

Flux linkage characteristics are discussed. Geometrical and performance specifications of a 3-phase 12/8 canned SRM used in this section are shown in Table 2.4. The can shield material is non-magnetic with high electrical resistivity. The rotor position 0° is defined as unaligned position while 22.5° is aligned position. The maximum DC phase current can reach up to 200A with number of turn per pole 10, allowing sufficient magnetic saturation. Flux curves are obtained by recording rotor positions at intervals of 3° increment within the entire range.

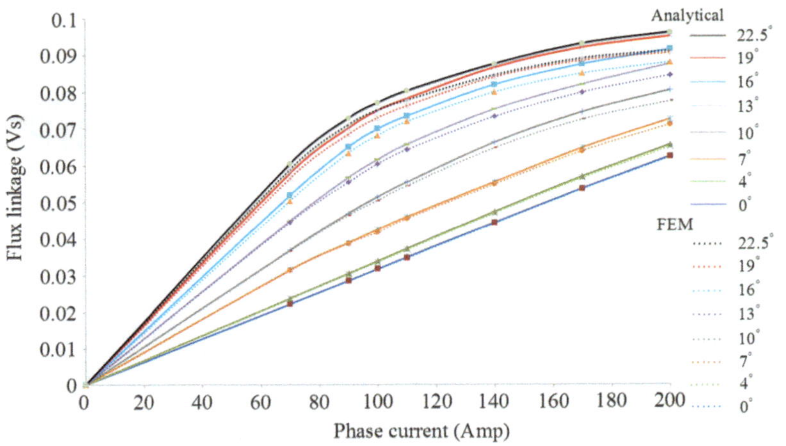

Fig. 2.14 Comparison of flux linkage characteristics of a canned machine

Numerical comparison of the curves of the canned SRM is shown in Fig. 2.14. Both the proposed analytical and FE method agree, especially when poles have none or little degree of alignment. With the degree of pole pair overlap increases, numerical gap occurs, and the FE values are lower. The gap is enlarged with higher phase current as well. The gap primarily occurs due to the prominent end effect for canned machines, which is explained as follows. When poles are unaligned and thus a very large equivalent airgap length, the low airgap flux will induce the weakened eddy current on cans and hence less the end effect. In such occasion, numerical agreement demonstrates high calculation accuracy. When poles have alignment and thus a sharply reduced airgap length, the airgap flux will induce strong eddy current that generates a circulation path on cans. At either axial endings on the can shield, obvious current concentration happens, causing strong electromagnetic interaction with airgap flux. Such interaction will consume magnetic energy, which is converted into ohmic loss on cans. Therefore, the airgap flux values are reduced close to pole alignment due to such end effect that is included in the 3D FE model while not in the proposed network. In addition, the more degree of alignment, the higher the numerical gap is. This is because of the reduced airgap reluctance that facilitates higher rate of change of airgap flux, causing stronger end effect.

The impact on flux curves by the use of can-shields is further illustrated in Fig. 2.15. By setting out electrical resistivity of both cans 0, the canned machine is reduced to an ordinary one of enlarged airgap. It is shown reduced values at each rotor position when cans are used, indicating magnetic energy consumption on cans. Numerical gap from both machines depends less on rotor position, but more on phase current levels. This is because for canned machine with lower phase current, the electromagnetic interaction on cans is weak, whereas with phase current increasing, the intensified airgap flux brings stronger electromagnetic coupling on cans.

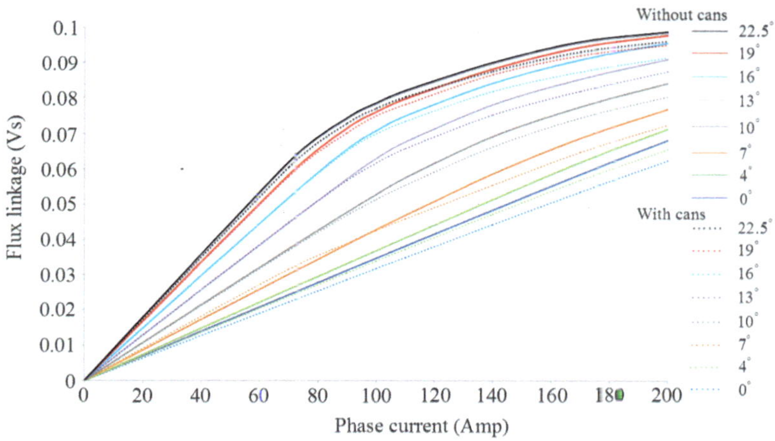

Fig. 2.15 Flux linkage curves and can effect comparison at selected rotor positons

2.3 Loss and Efficiency Analysis

In this section, an analytical method based on the circuit network model in Sect. 2.2 is proposed to estimate iron loss by non-sinusoidal flux excitation. The model is based on the definition of hysteresis and eddy current losses. By adding up the effect of minor BH loops, the accuracy of hysteresis loss is improved. The excessive loss, which is usually estimated using empirical equations, is not needed. Due to the use of limited number of discretized elements in the network, the computational effort is greatly reduced. Variable loss coefficients are applied according to derived magnetic flux densities in different parts of the machine.

2.3.1 The Calculation Method

The Steinmetz model is often used to estimate iron loss. The hysteresis loss P_{hys} and eddy current loss P_{edd} are calculated according to peak value of the flux density B and the frequency f.

$$P_{iron} = P_{hys} + P_{edd} = k_{hys} f (B)^{c_\beta} + k_{edd} f^2 (B)^2 \tag{2.1}$$

where k_{hys} and k_{edd} are hysteresis and eddy loss coefficients respectively. However, loss estimation in SRM using Steinmetz method might not be accurate, as the flux density and frequency are unevenly distributed in different locations on the core. It is difficult to estimate iron loss of the entire machine directly using a single expression of flux density as a function of time. The accuracy can be improved if the machine is discretized into elements and loss is calculated based on each of them.

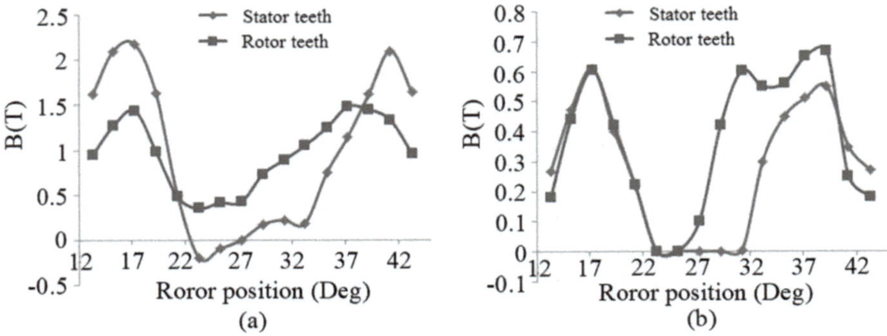

Fig. 2.16 Flux density variation of a single element on the stator and rotor teeth on different excitation modes for the studied 4-phase 16/12 traction SRM (**a**) low speed (**b**) high speed

Another problem of using Steinmetz model is that this calculation is only valid for sinusoidal flux variation. However, the flux waveforms in SRM are far from sinusoidal. The variation depends on rotor position and excitation mode. This can be observed in Fig. 2.16, where the flux waveforms as a function of time from the studied SRM have been shown at low and high speed operation.

At low speed (Fig. 2.16a), the curves start when the rotor and stator teeth are partially overlapping. The flux density increases with the degree of overlapping and then drops down sharply after the aligned position around 17°. When the teeth are not excited, the flux density remains very low. The flux on the stator may change direction due to mutual inductance. Similarly, the flux variation with continuous mode when the motor is running at high speed is shown in Fig. 2.16b. Likewise the waveform varies with rotor position. The flux density is reduced, as magnetic field is weakened by the back electromagnetic force.

When the flux density function is non-sinusoidal, Fourier method is used to sum up the loss from each harmonic [34]. For eddy current loss, this method is feasible. However, it is not applicable to hysteresis loss. This can be explained over the harmonics from a trapezoidal flux waveform in Fig. 2.17. The BH loop that each harmonic covers in the magnetization map is shown in Fig. 2.18. The dashed waveform represents the loop by the fundamental component while the remaining minor loops belong to higher order harmonics. The fundamental reflects the largest BH loop, while the minor loops have higher frequencies.

All the Fourier components together have a different BH loop and loss effect as compared to the trapezoidal waveform in Fig. 2.17. When all components are positive at a certain time, the loss generated is the sum of all harmonics. However, at a time instant where some components are negative (e.g. $t = 40$ ms in Fig. 2.17), the sum of each component using Steinmetz model will overestimate the loss, as both positive and negative components at this moment will separately contribute to loss.

One often used method to avoid overestimation of the hysteresis loss is to calculate the loss based on the fundamental waveform while adding another part

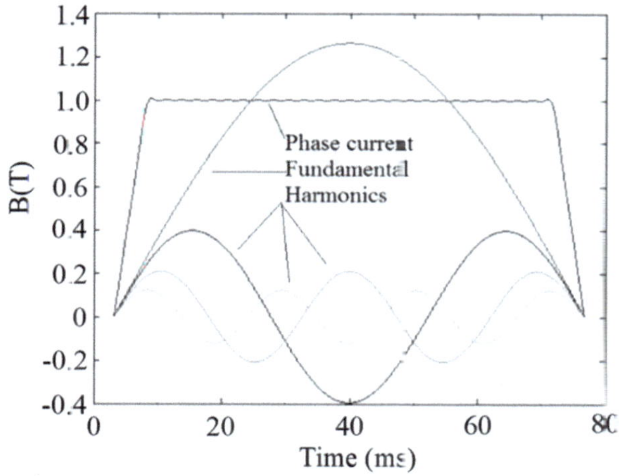

Fig. 2.17 The flux reversals by the fundamental and harmonics

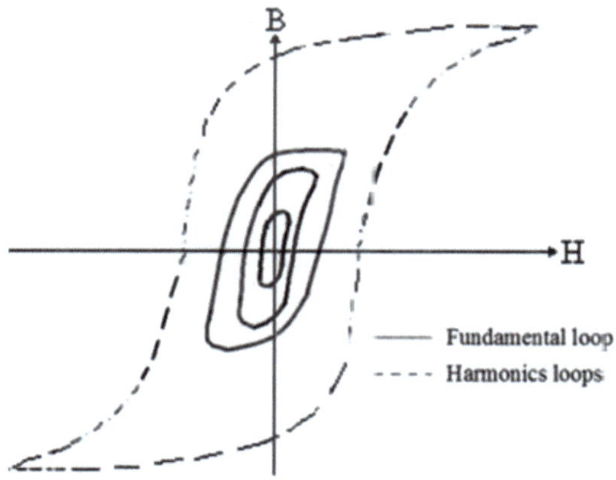

Fig. 2.18 The flux reversals by the fundamental and harmonics

representing excessive loss to estimate the magnetic particle friction [35, 36]. Another method follows a similar way and it takes the fundamental component into account, together with a correction factor [37]. However, these coefficients are complicated to obtain analytically.

In this method, the calculation of eddy current and hysteresis losses is based on the analytical definition of these loss components. As a result, the problems mentioned above are avoided. To account for flux distribution on different parts, the machine is analytically discretized and calculation is applied for each element.

2.3.1.1 Eddy Current Loss

For a single element i, the average loss density p_{edd_i} is presented as a function of square of the rate of change of flux density $B_i(t)$ as

$$p_{edd\_i} = \frac{1}{T_{int}} \int_0^{T_{int}} \frac{k_{edd}}{2\pi^2} \cdot \left(\frac{dB_i(t)}{dt} \right)^2 \cdot dt \tag{2.2}$$

where T_{int} is the time period and it equals to the time duration when rotor passes from aligned to unaligned position. Equation (2.2) can be discretized as

$$p_{edd\_i} = \frac{k_{edd}}{2\pi^2} \frac{1}{T_{int}} \sum_{k=1}^{N} \frac{(B_{k+1\_i} - B_{k\_i})^2}{\Delta t^2} \cdot \Delta t = \frac{k_{edd}}{2\pi^2} \frac{1}{T_{int} \Delta t} \sum_{k=1}^{N} (B_{k+1\_i} - B_{k\_i})^2 \tag{2.3}$$

where N is the number of steps during T_{int}. f is the frequency between aligned and unaligned position, Δt is time interval between simulation points t_k and t_{k+1} and is defined as

$$\Delta t = t_{k+1} - t_k = \frac{T_{int}}{N} = \frac{1}{fN} \tag{2.4}$$

$$k = 1, 2, \ldots\ldots, N \tag{2.5}$$

Combining (2.3), (2.4) and (2.5), p_{edd_i} is written as

$$p_{edd\_i} = \frac{k_{edd}}{2\pi^2} N_{step} f^2 \sum_{k=1}^{N} (B_{k+1\_i} - B_{k\_i})^2 \tag{2.6}$$

The eddy current loss by sum of elements M can be written as

$$P_{edd} = \sum_{i=1}^{M} m_i p_{edd\_i} = \frac{k_{edd}}{2\pi^2} N_{step} f^2 \sum_{i=1}^{M} \left(m_i \sum_{k=1}^{N} (B_{k+1\_i} - B_{k\_i})^2 \right) \tag{2.7}$$

where m_i is the mass of Element i.

2.3.1.2 Hysteresis Loss

The hysteresis loss is calculated based on peak flux density$\widehat{B}_i$. For a non-sinusoidal waveform, such assumption also applies, but only for the fundamental loop. The hysteresis loss density for the i-th element p_{hys_i} considering only the main loop is given as

$$p_{hys\_i} = k_{hys} f \left(\widehat{B}_i \right)^{c_\beta} \tag{2.8}$$

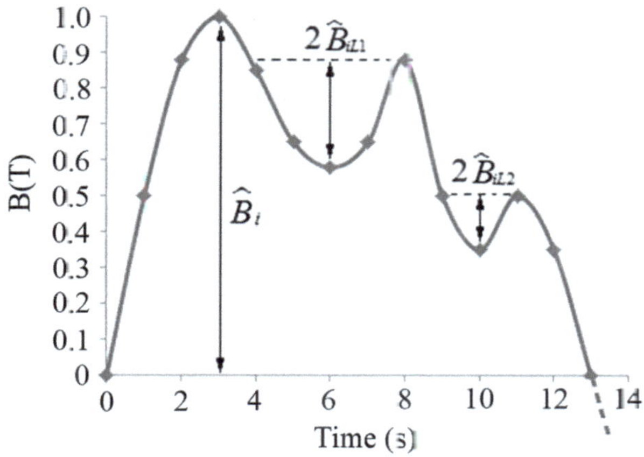

Fig. 2.19 The flux density waveform of an element showing the local flux reversals

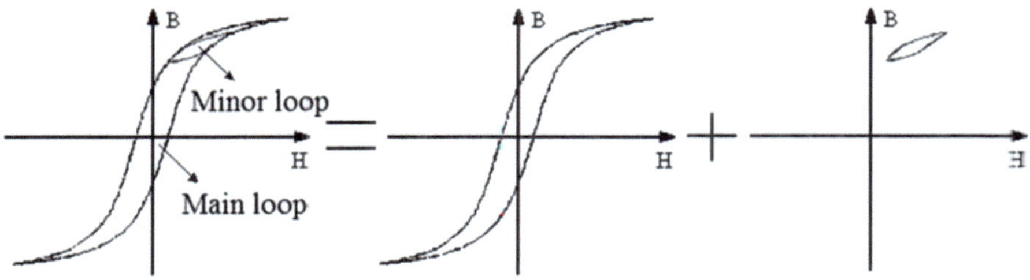

Fig. 2.20 Separation of reversals into main and minor loops

The flux density in (2.8) should be rewritten for minor loops by finding out the local peaks as depicted in Fig. 2.18. Only the positive flux density waveform is shown in this figure. It can be observed that there are two local peaks $\widehat{B}_{iL1}$and $\widehat{B}_{iL2}$.

In order to find the local peaks, an algorithm has been written in Matlab as a part of the proposed loss calculation method. Assuming that the flux density B_i at each time step t_k is known (Fig. 2.19), the global peak is detected by finding out the maximum value during the entire period, while local peaks are identified by comparing neighborhoods. If t_k is higher than both t_{k-1} and t_{k+1}, the local peak is detected. The local minimum is found in a similar way. $\widehat{B}_{iL1}$ and $\widehat{B}_{iL2}$ are determined by the difference between local peak and minimum values. The main loop and the minor loops are caused by global and local peaks, respectively. The hysteresis loss is calculated by sum of all the loops. The principle is illustrated in Fig. 2.20.

In (2.9), the hysteresis loss is calculated by the sum of all the elements M. For each element, the loss density $p_{hys\_i\_j}$ is calculated based on (2.8). The minor loops are also considered. N_{p_i} is the sum of flux density reversals of i-th element, and j is the counting number, $j = 1, 2, \ldots, N_{p\_i}$.

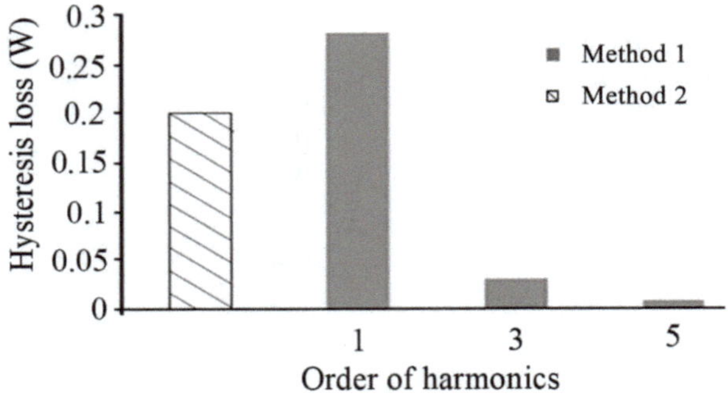

Fig. 2.21 Hysteresis loss by different methods

Table 2.5 Comparison of different methods

	Fourier	Direct
Hysteresis loss	0.3225	0.2019
Eddy loss	0.416	0.416

$$P_{hys} = \sum_{i=1}^{M}\left(m_i \sum_{j=1}^{N_{p\_i}} p_{hys\_i\_j}\right) = k_{hys} f \sum_{i=1}^{M}\left(m_i \sum_{j=1}^{N_{p\_i}} \left(\widehat{B}_{i\_j}\right)^{c_\beta}\right) \tag{2.9}$$

2.3.1.3 Comparison and Discussion

A comparison for the loss calculation between Steinmetz model (Method 1) and the proposed direct method (Method 2) has been provided using Maxwell FE software [38]. In Method 1, when flux density is non-sinusoidal, Fourier components are used, while Method 2 is based on (2.7) and (2.9). The loss coefficients are set as $k_{edd} = 8.143 \times 10^{-5}$ W/kg, $c_\beta = 2$ and the fundamental frequency is 2.4 kHz. For the comparison, flux density waveform of an element on the stator teeth in Fig. 2.16a is used. The results are shown in Fig. 2.21. The hysteresis loss from the fundamental frequency takes up the main portion, while losses from the harmonics are much less. It is also shown that the loss by the direct method is lower than loss from the fundamental frequency using Steinmetz model.

Table 2.5 summarizes the loss values calculated by each method. It can be observed that eddy current losses are the same in both methods while hysteresis loss by Fourier method is higher. As discussed previously, hysteresis loss is overestimated in Steinmetz model which uses the sum of Fourier components. In the Sect. V, the accuracy of the proposed method will be provided by a comparison with FE analysis results.

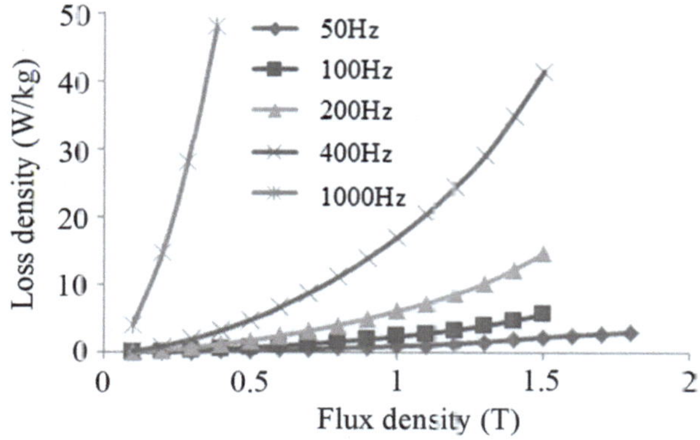

Fig. 2.22 The loss curves of M1529G

2.3.2 *The Variable Loss Coefficients*

The frequency and flux density values, which the hysteresis and eddy current loss coefficients k_{edd} and k_{hys} are linked with, depend on where they reside and operating condition of the machine. It is not accurate to calculate the entire loss of the machine using one constant coefficient. In the proposed method, loss coefficients are obtained, exclusively for each element, based on loss characteristics of the laminations.

As to how to select lamination, the commonly used M-cores series and the JNEX super core are considered. For a traction application, M-cores offer lower prices, while JNEX core offer lower core losses since it is only 0.1 mm thick. However, it is much more expensive than M-cores. M1529G and M1929G are typical categories of M-cores. Both have the same thickness of 0.35 mm. M1529G has lower hysteresis loss coefficients. The eddy current coefficients of these two materials are the same.

It is necessary to make detailed loss analysis before making a prototype. Figure 2.22 shows loss and flux density relationship of M1529G. The loss increases geometrically with higher flux density and frequency. The loss above 0.5 T is not directly provided under 1 kHz and higher. As a result, the corresponding coefficients cannot be directly derived. The coefficients are derived by using curve fitting method. Using $c_\beta = 2$, (2.1) can be reorganized as

$$\frac{P_{iron}}{f\widehat{B}^2} = k_{hys}(f, B) + k_{edd}(f, B)f \tag{2.10}$$

The relation between loss and frequency is modified and shown in Fig. 2.23. The group of curves is highly linear. The slope of curves indicates the value of k_{edd} and the length on the vertical axis shows k_{hys}. The coefficients are considered to depend on flux density only.

k_{edd} and k_{hys} are expressed using a cubic polynomial, they are written as

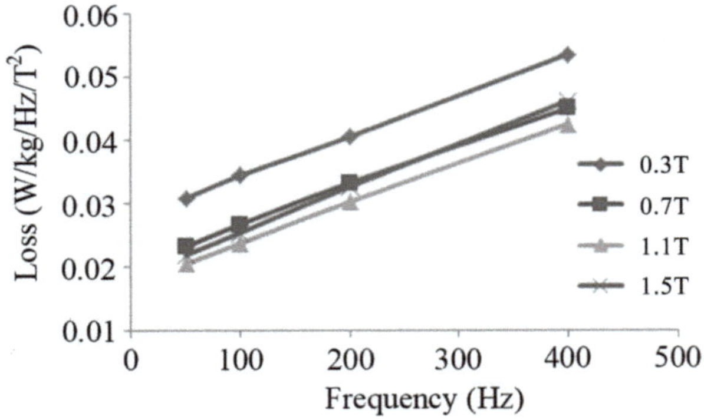

Fig. 2.23 Linearized loss curves

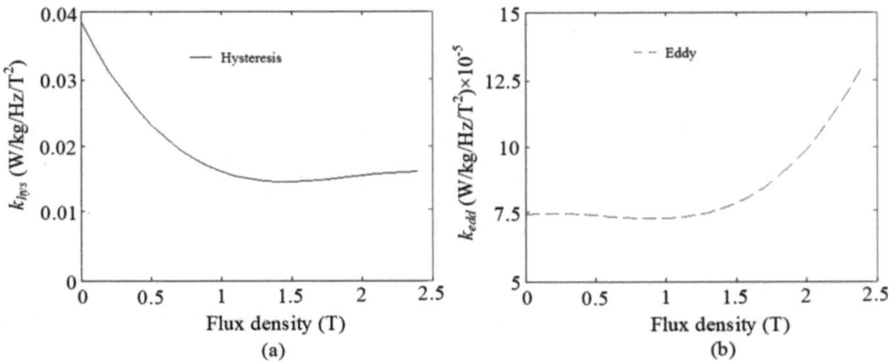

Fig. 2.24 Loss coefficients with flux density, (**a**) hysteresis, (**b**) eddy

$$k_{edd} = k_1 + k_2B + k_3B^2 + k_4B^3 \tag{2.11}$$

$$k_{hys} = k_5 + k_6B + k_7B^2 + k_8B^3 \tag{2.12}$$

To get k_1–k_4 in (2.11), 4 equations are needed which are obtained from the slopes of the 4 lines in Fig. 2.23. Similarly k_5–k_8 are obtained from the values intersecting with vertical axis.

The coefficients as a function of flux density are shown in Fig. 2.24. The hysteresis loss coefficient drops down with flux density till 1 T, showing that it becomes difficult to get the magnetic domain further biased. On the contrary, the eddy current coefficient rises up from about 1 T. In this method, the coefficients k_{edd} and k_{hys} are obtained for each element based on flux density only. For this purpose, a look-up table based on Fig. 2.24 is used, as part of the loss calculation algorithm.

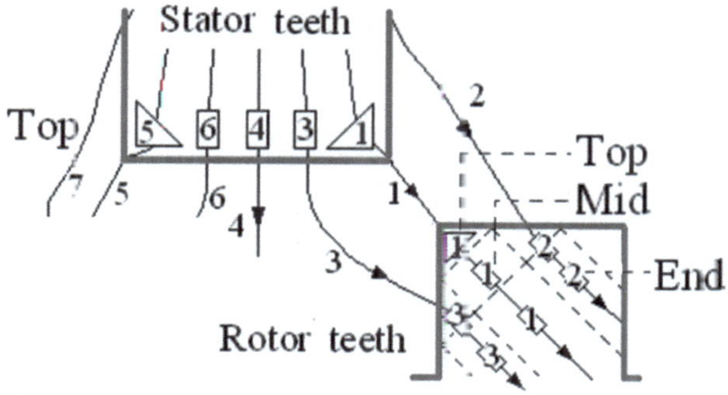

Fig. 2.25 The magnetic equivalent paths when teeth are not aligned

2.3.3 *The Discretized Elements*

For a loss analysis, the flux density of each element has to be calculated as a function of time. One way of obtaining the flux densities is to obtain them through FE analysis. For this purpose, at each rotor position, the flux density of each element is stored in a matrix, with its position recorded. The change of flux density over time ($B_{k+1\_i}-B_{k\_i}$) is arithmetically calculated from each data in the matrix. This method is accurate but time consuming.

In this loss calculation method, a simple analytical magnetic equivalent circuit method described in Sect. 2.2 is used to simplify the use of elements. The flux distribution in the air gap is the key feature. To describe how the magnetic flux passes through the air gap, the equivalent circuit is divided into paths. Figure 2.25 shows the paths at non-overlapping rotor positions where poles have no alignment. There are 7 paths when stator and rotor teeth are not aligned. Paths 1–3 account for the main ways that the flux goes into the rotor teeth while Path 4 is for slot leakage. Paths 5–7 are similar to Paths 1–3. When teeth begin to overlap, Paths 5–7 are not needed. The size including length and width of the air gap reluctance of each path is calculated according to rotor position and geometry with curvature considered, and an equivalent circuit is created to calculate the magnetic flux densities. For each path, the teeth and yokes are further divided into components as top, mid and end to account for different degree of saturation. The size of each reluctance component is adjustable according to rotor position and excitation level.

The effect of nonlinear magnetization characteristics of the lamination material have been taken into account by applying iterative algorithms. As the flux density in SRM is unevenly distributed, a 2-layer loop structure is used (Fig. 2.26) to increase the calculation accuracy. Starting from an initial permeability value for each reluctance component, inner loop is satisfied by convergence of permeability. To coordinate different paths, the outer loop is applied based on the principle that the flux density on the yoke for each path should converge. Compared with FE analysis, the number of elements using this method is reduced significantly.

An accurate phase current waveform is a prerequisite for loss analysis and it depends on the dynamic phase inductance $L(\alpha,i_{ph})$ which should be calculated in a

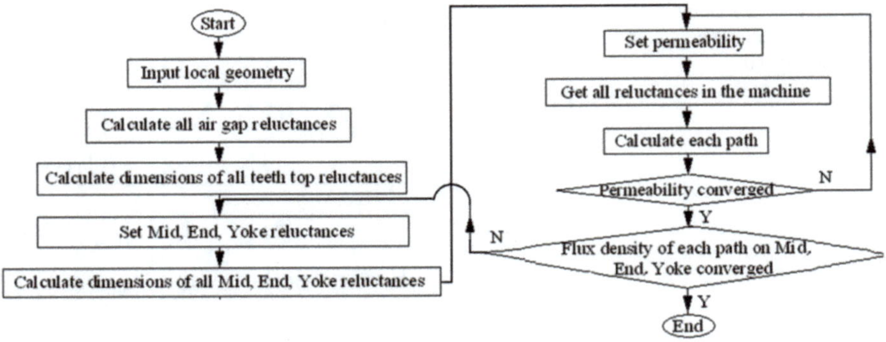

Fig. 2.26 The flow chart to get all elements in the proposed method

Fig. 2.27 The phase inductance characteristics showing variation of the air gap plus saturation using the proposed magnetic circuit method

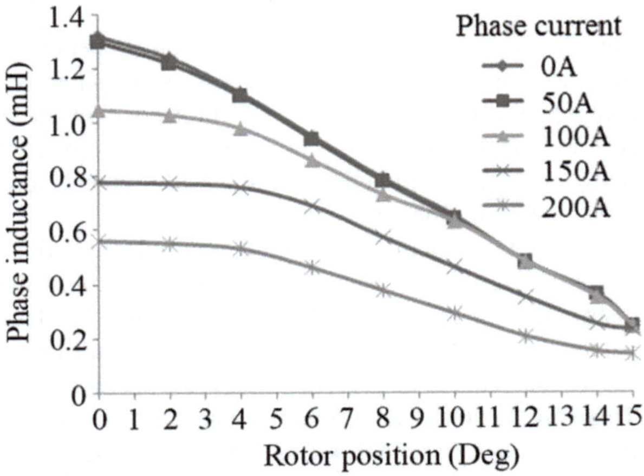

fast and accurate way. $L(\alpha,i_{ph})$ has relations with rotor position α as well as saturation due to the phase current i_{ph}. $L(\alpha,i_{ph})$ is obtained by the proposed magnetic circuit. The inductance of each magnetic path L_i ($i = 1,2,\ldots,N_{mag}$, where N_{mag} is the total number of paths) in Fig. 2.25 is first calculated and then $L(\alpha,i_{ph})$ is obtained by the followings.

$$L(\alpha, i_{ph}) = \sum_{i=1}^{N_{mag}} L_i(\alpha, i_{ph}) \tag{2.13}$$

In order to calculate the phase inductance, an algorithm has been created in Matlab. The phase inductance varies with current and rotor position using the magnetic circuit is shown in Fig. 2.27. It drops down from aligned 0 to the unaligned position 15° due to the increased air gap. For each rotor position, the value is decreased with higher current, showing the saturation effect especially nearby aligned positions. A dynamic half bridge asymmetrical controller is built up to get

Table 2.6 SRM volumetric constraints and features

Outer Diameter (mm)	264	Air gap (mm)	0.5
Axial length (mm)	108	Max. stator temperature	150 °C

Table 2.7 Local flux density (T) at saturated regions under phase current 200A

	ST.tip		*RT.tip*		*RT.mid*	
Rotor Angle (Deg)	I	II	I	II	I	II
1	2.31	2.29	2.05	2.11	1.8	1.85
5	2.33	2.37	2.13	2.19	1.66	1.82
10	2.37	2.26	2.22	2.0	0.82	0.73
14	0.75	0.77	0.58	0.52	0.47	0.41

the dynamic phase current waveform. $L(\alpha,i_{ph})$ from Fig. 2.27 is used in the form of look-up table to model the circuit.

2.3.4 Verification and Discussion

The proposed method has been validated on a 4-phase 16/12 SRM for traction application. The overall geometry is shown in Table 2.6. Each coil has 10 turns. Fewer turns are preferred to have a wide speed range capability. To have a high starting torque and power density, high phase current up to 240A is applied, which causes high local saturation in the machine.

Table 2.7 shows the flux densities of typical elements such as on the tip stator teeth (*ST.tip*), tip of the rotor teeth (*RT.tip*) and middle of the rotor teeth (*RT.mid*). For the mechanical angles, 1° of rotor angle represents the closest position to the aligned position when phase current begins to drop down, while 14° is the closest position to the unaligned position where phase current is about to rise up. The peak flux density is observed at the tip of the poles, especially when teeth are overlapping. As is shown, there is agreement between FE results and the analytical method. Since the equivalent network calculates the flux densities in a faster way, the proposed method is further used to get the flux density variation on different part of the machine.

For this purpose, an analytical algorithm using Matlab is designed and its flow chart is shown in Fig. 2.28. The first step is to get the geometry and flux density of each element at all time-steps. In the magnetic circuit calculations, the time interval Δt is selected small enough to recognize the slight difference of flux when rotor position changes. The phase current waveform is obtained with variation of phase inductance $L(\alpha,i_{ph})$. Then, the hysteresis and eddy current losses are calculated separately. The hysteresis loss by both the main loop and local peaks are considered. The eddy current loss is calculated according to the variation of flux density between each step. During the iterations, the variable coefficients in Fig. 2.24 are used.

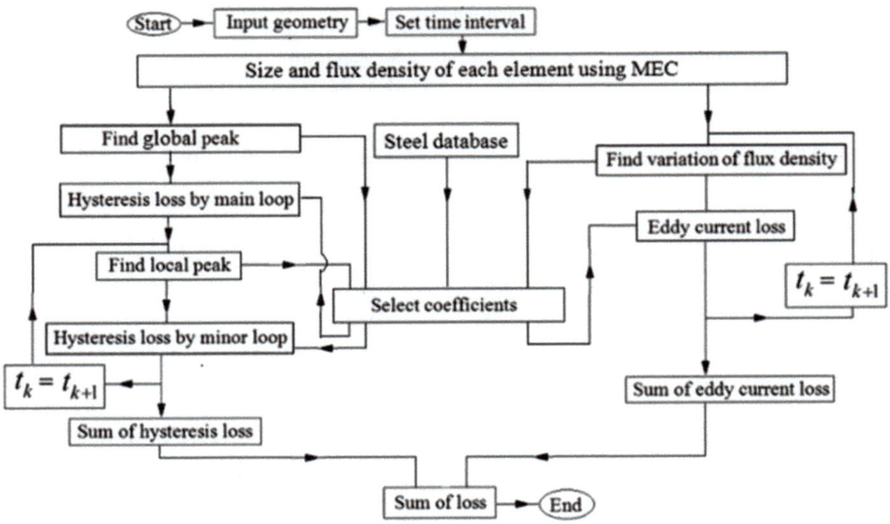

Fig. 2.28 The flow chart of the program

Fig. 2.29 The current-loss relationship from different methods at low speed

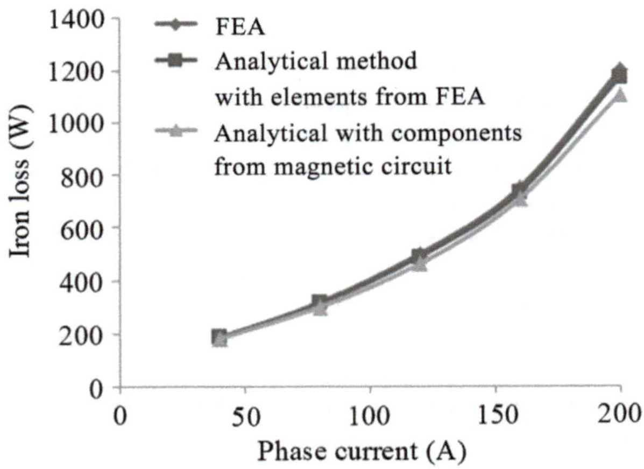

Iron loss under different excitation levels is shown in Fig. 2.29, where both FE and the analytical method are used. For the proposed method, elements both from FE analysis and the magnetic circuit method are utilized. The loss from FE and the proposed method with elements from FE analysis fit each other well. However, getting smaller elements in FE analysis requires significant computation time. The elements obtained from the magnetic circuit are simpler and much less in number. However, this reduces the accuracy of the proposed method, especially at high excitation current, when high level of local saturation occurs. In this case, the limited number of elements could not cover exactly the fast variation of the flux density with geometry. Such error can be improved by making elements finer in the magnetic

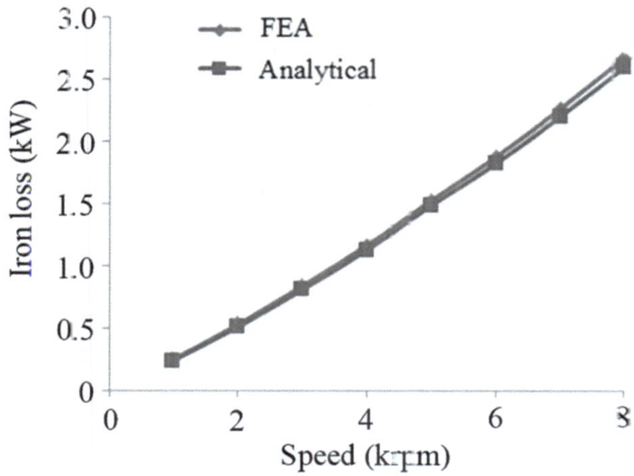

Fig. 2.30 The speed loss relationship from different methods

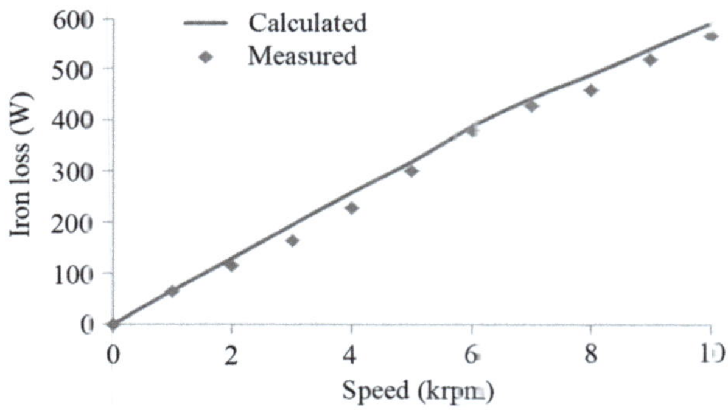

Fig. 2.31 Comparison between measured and predicted losses under no saturation condition (constant torque 5 Nm)

circuit, but at cost of computation time. In most cases, a simple discretization is sufficient for a wide power range.

The method is also applicable to a wide range of rotating speed. As shown in Fig. 2.30, the loss speed relationship under constant current (150A) excitation, both from FE analysis and the proposed method, shows good agreement. This method has no limitation on rotating speed or power level. The SRM geometry can be easily changed and the sizes of discretized elements should be modified accordingly. The method is temperature independent as loss coefficients of lamination rely on frequency only.

The proposed method is also verified through experiments on a 30 kW SRM which is mechanically connected to a DC machine dynamometer. The dSPACE 1103 system is used to connect the SRM and converter. Figures 2.31 and 2.32 show the comparison between the measured and calculated losses under constant torque or speed condition. It shows obviously a good agreement. The error of the predicted loss is less than 10%. The most deviation occurs at high torque when the machine is greatly saturated, due to the limited number of elements in the proposed magnetic circuit.

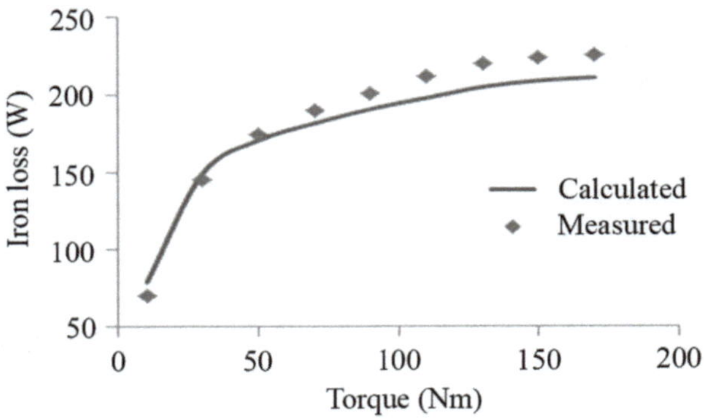

Fig. 2.32 Comparison between measured and predicted losses with saturation, with constant low speed 1000 rpm

2.4 A Simplified Network Model

In this section, the proposed circuit network model is further discussed. Modifications include the application of simplified airgap paths, a lookup method for magnetic saturation estimation, as well as considerations of magnetic coupling between phases. The airgap reluctance is modeled by 3 representative paths, one principal and a coupling of in-profile ones that together form the fundamental network structure. Further to account for locally distributed magnetic saturation, a fast FE lookup method is proposed, which is applied to geometry variation of the machine without power limit. The amount of computation is quite competitive compared with FE.

The magnetic phase coupling occurs close to the full alignment where a couple of phase is excited simultaneously. Taking such a point can improve calculation accuracy [39] in flux and torque control. However, most magnetic equivalent circuit (MEC) models do not take into account the coupling and only a few are involved [40, 41]. In [40], the coupling effect is taken by adding equivalent airgap reluctance between phases. The elements values can be predetermined by FE analysis or measurement. However, the calculation accuracy needs further discussion if taking a closer comparison with FE, especially close to the aligned rotor position. The proposed network in this section takes such magnetic coupling between phases.

2.4.1 The MEC-FE Model

Fig. 2.33 shows flux paths at typical rotor conditions. In Fig. 2.33a, one phase at unaligned rotor position is excited, the magnetic circuit is unlikely to get saturated and airgap reluctance is considered as the main factor. In Fig. 2.33b, a couple of pole pairs, including the one of defluxing close to the full aligned position while the other pair of fluxing just after the unaligned position, are energized simultaneously. The multi-phase excitation may lead to not only magnetic saturation, but also the magnetic coupling between phases. Therefore, phases are related and the degree of

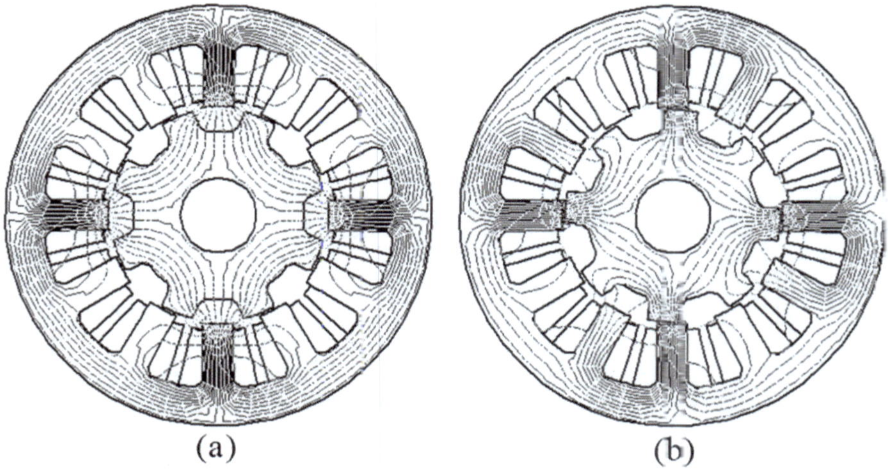

Fig. 2.33 Flux paths overview at typical rotor positions, (**a**) at unaligned position by one single phase excitation, (**b**) at partial aligned position by two phase excitations, showing magnetic coupling effect

Table 2.8 Geometry overview of the studied SRM

Stator bore outer radius (mm)	67.5	Stator pole width (deg)	15.5
Stator bore inner radius (mm)	34.85	Rotor pole width (deg)	17.5
Rotor bore outer radius (mm)	33.95	Airgap length (mm)	0.9
Rotor bore inner radius (mm)	27	Stack length (mm)	125
Shaft radius (mm)	12	Stator yoke thickness (mm)	12

dependence is deemed as a function of rotor speed and phase current. Table 2.8 shows overview of the studied 3-phase 12/8 SRM according to Fig. 2.33.

The proposed MEC-FE model is shown in Fig. 2.34. The stator and rotor are divided into sectors and each is represented by a lumped magnetic reluctance element, including stator yoke (SY), stator tooth (ST), rotor tooth (RT), rotor yoke (RY). Compared with the discretized model, the number of elements on poles is reduced to the minimum numbers. However due to magnetic saturation that predominantly resides at the exciting poles, elements on the stator and rotor teeth are marked with arrows, indicating variability to be determined by a lookup approach. Each rotor slot is represented by an element LK short for slot leakage.

The airgap reluctance (AG) is defined as magnetic connection between a stator and rotor pole pair. The connection is valid from the aligned position till the position when the adjacent pole pair is aligned. That is for one pole pair, the minimum airgap reluctance occurs at its alignment while the maximum occurs at the position when the adjacent pair is aligned. In such a way the regarded system matrix is composed of reluctances from all poles and particularly includes the airgap connections from stator to rotor poles. Therefore in Fig. 2.34, each stator pole has a couple of

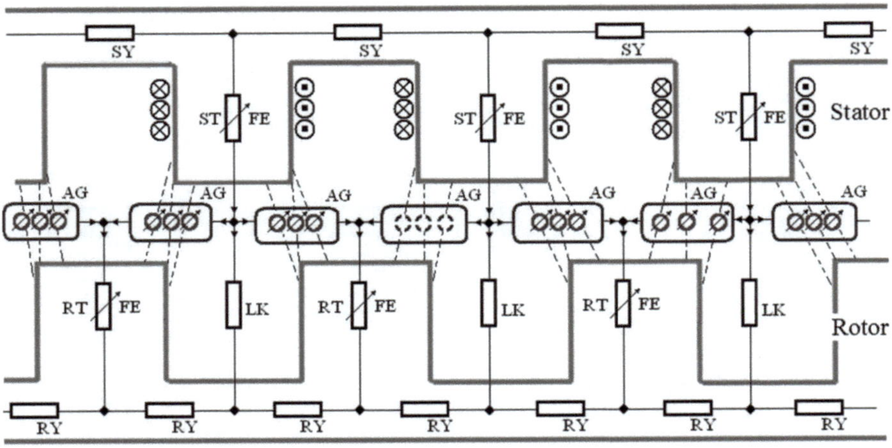

Fig. 2.34 Overview of the proposed network model

Fig. 2.35 Airgap flux path details, showing the non-overlap and partial overlap of a pole pair

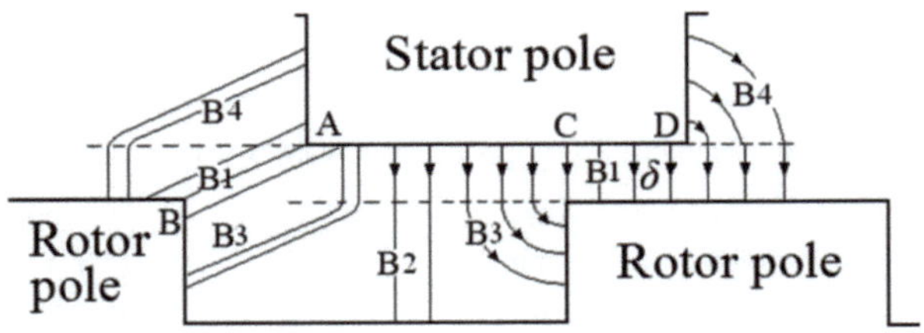

connections to the previous and next rotor poles. Further for each connection, 3 paralleled branches are shown. The middle branch stands for the tip-to-tip path while both side branches stand for the in-profile paths. For each branch the equivalent reluctance is represented by a ring, and the arrow indicates variability.

Figure 2.35 shows the airgap flux path details for the airgap reluctance rings, with typical positons of both non-overlap and partial overlap. In either case, the airgap paths are modeled by 3 branches B1–B3–B4 and each is represented by a lumped magnetic reluctance element. The reluctance value can be calculated according to the length and width of the regarded airgap path. When poles are not aligned, B1 refers to the shortest distance between the stator and rotor pole tips. The length of B1 is determined by such shortest distance while the width is by a fitting approach in Fig. 2.2. B3 refers to the side path through rotor slot. The length is determined by the straights both in airgap and rotor slot while the width is by comparing with slot leakage path B2 according to the magnetic circuit principle. B4 path can be seen symmetrical to B3. When poles are partially aligned, B1 refers to the overlap airgap zone and the path is determined by airgap length δ and overlap width CD. In analogy B3 is calculated, to which B4 is likewise symmetrical.

Besides airgap, magnetic saturation as another source of reluctance affects flux linkage curves. The saturation serves as a function of phase current excitation, as well as rotor position that depend on airgap path. It is assumed that the degree of saturation is quantifiable and this paper uses a lookup approach to identify the value in the form of the permeance drop λ, the reverse of reluctance. Due to double saliency, the lookup is separated according to whether the pole pair has alignment. The drop is described as the reduced permeance value in the MEC network if both the ideally non-saturated and actually saturated electrical steels are respectively applied, provided that the operation conditions are the same.

To make the lookup for magnetic saturation dimensionless, how to model the rotor positon with minimum variables is described. The saturation quantity in value is independent of a specific machine and suited to geometry variation. Not only the position information of a pole pair takes effect, but also the level of phase current is required. Therefore instead of using curves, a fitting surface as a function of position and phase current is proposed. The position regards to the relative location of a pole pair being excited, to which the airgap length, widths of poles and rotor positions may contribute. These factors will take effect in different ways according to whether that pole pair has any degree of alignment. Accordingly, the fitting is separately modeled whether poles have degree of overlap.

When poles are not aligned, rotor position is represented by the length AB in Fig. 2.35 that the main part of flux lines is going through airgap. Here it is defined as distance L_{ab} that integrates rotor angular shift θ and radial airgap length δ by the mentioned fitting approach. Then, permeance drop is defined as $\lambda(i_{ph}, L_{ab})$, where i_{ph} donates phase current excitation. Let $L_\theta = L_{ab}$, $\lambda(i_{ph}, L_{ab}) = \lambda(i_{ph}, L_\theta)$ holds true. When poles have alignment, Fig. 2.35 shows a featured horizontal overlap CD as main flux paths. In this case, the overlap width L_{cd} and airgap length δ are defined. According to permeance calculation $\Lambda_m = (\mu{\cdot}S) / L$, where L, S are respectively length and width of magnetic path, we define the airgap length per unit overlap width δ / L_{cd}, which in analogy integrates the radial airgap length δ, as well as the rotor angular shift θ in the form of the overlap length L_{cd}. Let $L_\theta = L_{bc}$, the permeance drop $\lambda(i_{ph}, L_{bc}) = \lambda(i_{ph}, L_\theta)$ holds true.

Now for both cases, the permeance drop $\lambda(i_{ph}, L_\theta)$ is deemed as a function of phase current i_{ph}, as well as relative position of a pole pair L_θ that measures airgap reluctance variation. Further to make the lookup independent of a specific machine geometry, the permeance drop lookup function $\lambda(i_{ph}, L_\theta)$ is transformed. Let $\lambda^* = \lambda\ L_z / \mu_0$, meaning the unit permeance drop in vacuum, where L_z is the axial length and μ_0 is permeability of air. Now the lookup function is $\lambda^*(i_{ph}, L_\theta)$.

The permeance drop is modeled by a 3D lookup. Figure 2.36 shows the map at non-overlap and overlap rotor positions respectively. Note that the lookup applies to the machine with number of turns per pole $N_{tp} = 10$, and if N_{tp} differs, the phase current varies accordingly in reversely proportion, as long as the ampere-turn value keeps constant. In Fig. 2.36a, λ^* value rises up with the increased phase current i_{ph}, as well as the decreased rotor position L_θ. When L_θ closes to 0 that poles are about to overlap, λ^* increases sharply with i_{ph}. This is because of the low L_θ value that is comparable to airgap length δ, making λ^* variation sensitive. In Fig. 2.36b, λ^* value rises up with both i_{ph} and L_θ. The increased L_θ means more degree of overlap till full

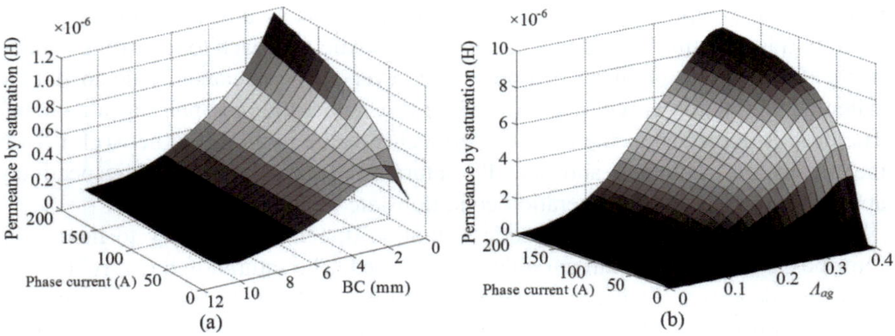

Fig. 2.36 The permeance drop lookup function obtained from FE, (**a**) non-overlapping; (**b**) partial and full overlapping

Fig. 2.37 The flux linkage under selected rotor positions without magnetic coupling

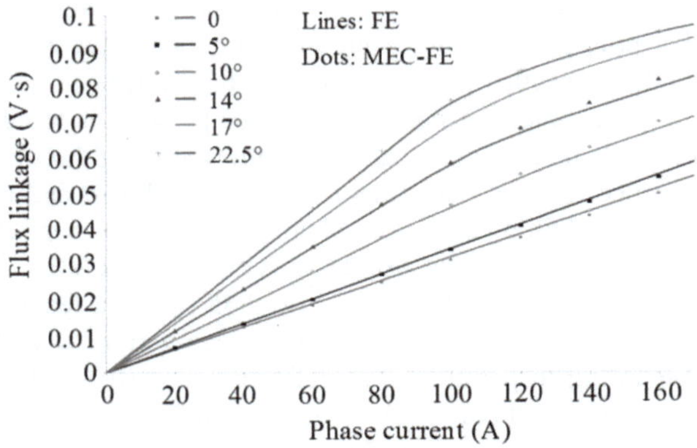

alignment, where the λ^* value rises up with i_{ph} in a saturated way, indicating strong magnetic saturation on poles.

The flux linkage curves of the studied machine by one single phase excitation at selected rotor positions are comparatively shown in Fig. 2.37, with a wide current range till 170A. Note that 0° is the unaligned rotor position while 22.5° is the aligned position. The selected positions include the total unaligned 0°, partial unaligned 5°, just before the start of overlap 10°, just after the start of overlap 14°, wide overlap 17° and total aligned 22.5°. Each curve by the proposed network model in general agrees with FE method. At unaligned position 0° where magnetic saturation unlikely happens, both methods agree, which proves high calculation accuracy of airgap reluctance. At the total aligned position 22.5°, the airgap path calculation is simple, and the agreement indicates high lookup accuracy of magnetic saturation. As to 10° and 14°, there exists a numerical discrepancy between FE and the network model with increase of phase current. When close to the start of overlap, high local magnetic saturation occurs on tips of poles, making the lookup less accurate. This is because of the simplification by integrating rotor position θ and airgap radial length δ into one variable in the lookup table. When poles are far unaligned or widely

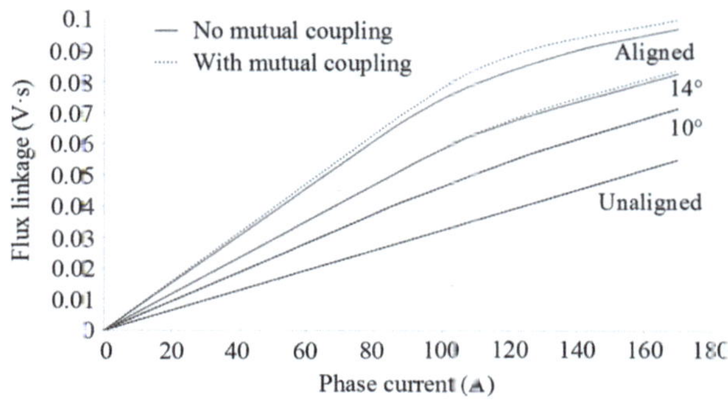

Fig. 2.38 Comparison of flux linkage curves under selected rotor positions, with and without magnetic coupling between phases

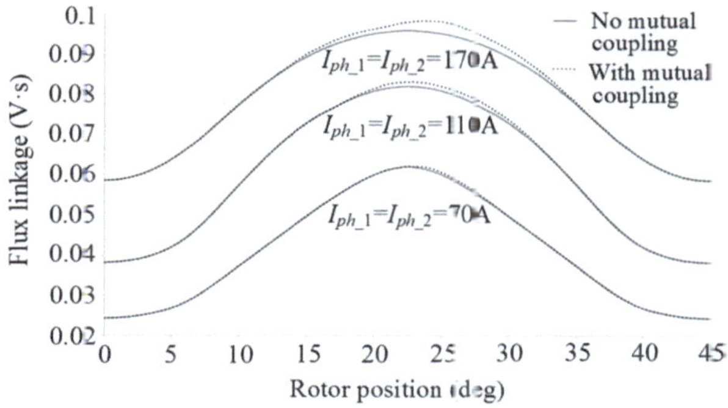

Fig. 2.39 Comparison of flux linkage variation due to the coupling effect, where different levels of current excitation are applied

aligned (5° and 17°), such numerical discrepancy reduces, as magnetic saturation either doesn't exist or is more evenly distributed on poles. In most cases, this network model is sufficient for a wide power range.

Further Fig. 2.38 comparatively shows magnetic coupling effect at selected rotor positions. At unaligned 0° and 10° where only one phase is working, according to the operation principle in Fig. 2.33a, both curves are exactly identical. However when poles are further approaching (at 14° and aligned 22.5°), magnetic coupling is involved by a couple of phase excitations, as is shown in Fig. 2.33b. In this case, both phases are loaded with current 170A. It is shown that the numerical gap of the curves at 14° or aligned 22.5° whether the coupling effect is taken is enlarged with increase of current. Also the numerical gap is more enlarged at 22.5°, indicating stronger coupling effect.

The magnetic coupling is further discussed. Figure 2.39 shows flux linkage variation under full rotor position range at selected levels of phase current I_{ph_1} and I_{ph_2} for a couple of phases. Phase 1 is the main phase for flux linkage calculation while Phase 2 is the auxiliary coupling phase, where 22.5° is the aligned position for Phase 1. According to the operation principle, the coupling mainly occurs at both sides of 22.5° where both phases are excited simultaneously. It is shown that (1) For each current level in Phase 1, the highest numerical discrepancy between the solid and dotted curves occurs at both sides of 22.5°, to which the dotted curve with coupling considered is not symmetrical. This is because before 22.5°,

Table 2.9 Geometric variation (in %) from the studied SRM in Table 2.8 as M0

	β_s	β_r	δ	l_{sy}
M0	100%	100%	100%	100%
M1			150%	
M2			70%	
M3		80%		
M4	80%	80%		
M5				75%

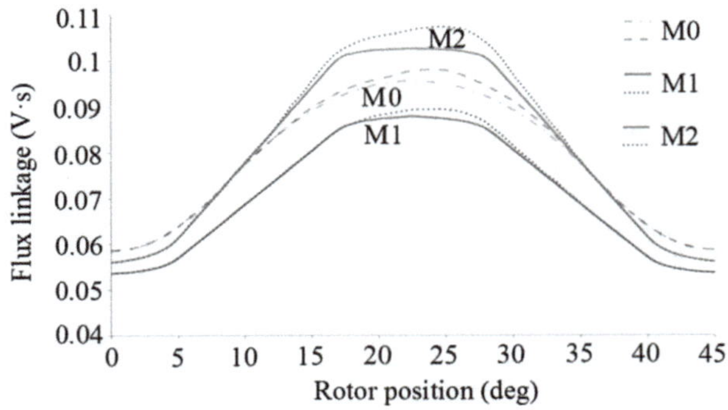

Fig. 2.40 The magnetic coupling effect due to variation of airgap length, where M1 and M2 are simulated, with M0 as reference; Note that 22.5° is the aligned position and phase current 170A is applied; Dots: with phase coupling; Lines: no phase coupling

Phase 2 is has less degree of alignment and contributes less on the coupling flux while thereafter, this phase has more alignment that in turn strengthens the coupling flux. (2) Higher current excitation causes stronger coupling effect.

The model adaptability is verified by simulating typical SRMs of the main geometric parameter variation. Table 2.9 shows the variation, including airgap length δ, stator pole width β_s, rotor pole width β_r and stator yoke thickness l_{sy}, based on the studied machine in Table 2.8 as M0. M1 and M2 are respectively defined as the enlarged and reduced airgap length variation. M3 and M4 are the reduced pole widths. M5 is the reduced stator yoke thickness. In the following, each machine is simulated with and without magnetic coupling.

Figure 2.40 shows flux linkage due to variation of airgap length (M1 and M2). The values decrease with a higher airgap length, especially close to 22.5°. Regarding the magnetic coupling effect, the reduced gap length of M2 strengthens both the main and coupling flux, leading to a higher discrepancy value between the solid and dotted curves compared with M0 or M1.

Figure 2.41 shows flux linkage due to variation of pole width. In M3, width of rotor pole is reduced by 20%, followed by M4 that the stator pole width is further reduced by 20%. It is found that the values at any rotor position are reduced in M3, and further sharply reduced in M4. Also, the peak width of the curves is narrowed with reduced pole width. However for each machine M0, M3 and M4, the numerical discrepancy due to magnetic coupling effect remains similar. This is because nearby the aligned positon 22.5° where the coupling takes principal effect, variation of pole widths hardly links with degree of magnetic saturation.

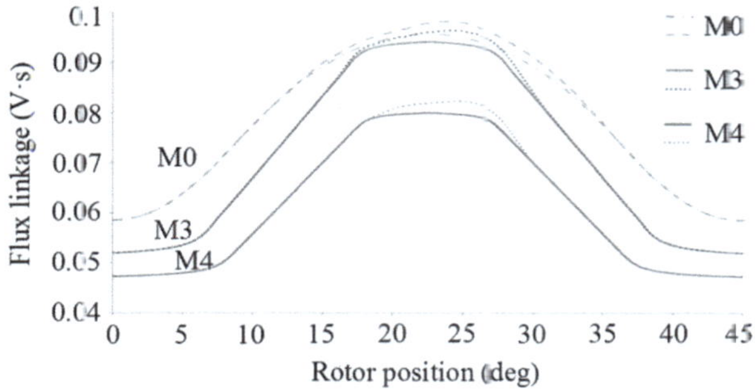

Fig. 2.41 The magnetic coupling due to variation of pole widths where M3 and M4 are simulated, with M0 as reference; Note that 22.5° is the aligned position and phase current 170A is applied; Dots: with phase coupling; Lines: no phase coupling

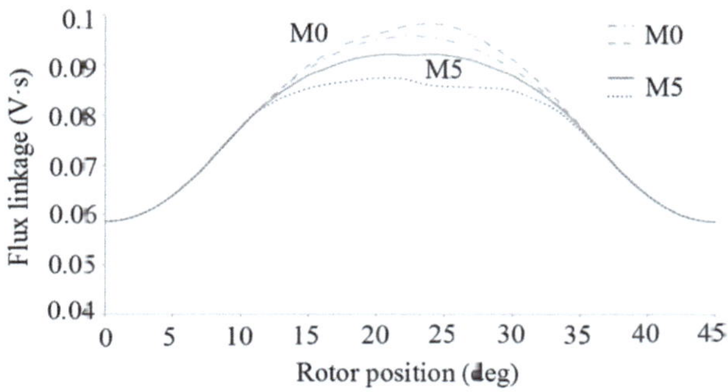

Fig. 2.42 The magnetic coupling effect due to stator yoke widths where M5 is simulated, with M0 as reference; Note that 22.5° is the aligned position and phase current 170A is applied; Dots: with phase coupling; Lines: no phase coupling

Figure 2.42 shows flux variation due to stator yoke thickness. It is found that M5 values by reducing 25% yoke thickness, is reduced. This is due to the narrowed yoke that leads to magnetic saturation that weakens active flux. Regarding the magnetic coupling effect in M5, discrepancy happens in particular around 22.5°, where the auxiliary coupling phase (Phase 2) takes principal effect. This is because of the coupling phase that intensifies saturation in the narrowed stator yoke.

In summary, this section proposes a FE assisted magnetic circuit model for SRMs, including magnetic phase coupling effect. The airgap reluctance is modeled by 3 typical paths without empirical estimation. To account for local magnetic saturation, a fast FE lookup is proposed, which can be applied to variation of the machine geometry without power limit. The amount of computation is quite competitive. Further this model can effectively simulate magnetic coupling effect between phases. It is shown significance the magnetic coupling affects flux and therefore the design can be better predicted under various control modes.

2.5 A Fitting Method for Airgap Reluctance

In this section, the network circuit in Fig. 2.34 is further discussed by alternatively modeling of airgap reluctance using a fitting approach. The proposed MEC is shown in Fig. 2.43 that is featured by nodal magnetic potential values In analogy, each

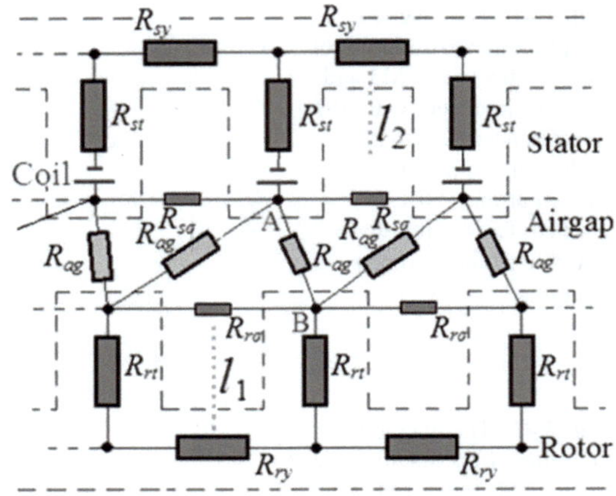

Fig. 2.43 The proposed MEC network with a fitting method for airgap reluctance

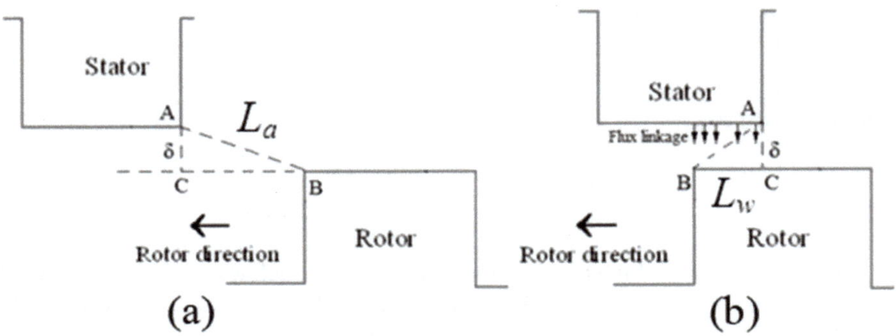

Fig. 2.44 Geometrical definition of positions for a pole pair, (**a**) non-overlap, (**b**) partial overlap

tooth is presented by a lumped reluctance, cutting yokes into sectors and each is likewise lumped. The permeability of iron is initially set infinite against that of air. There are N_s stator poles and N_r rotor poles and each is lumped with a magnetic reluctance R_{st} or R_{rt} respectively. The stator and rotor yokes are segmented and each sector is represented with R_{sy} or R_{ry}. $R_{s\delta}$ and $R_{r\delta}$ connect poles, serving as slot leakage. The magneto-motive force (MMF) is applied on one phase. There are altogether $(N_s + N_r)$ flux loops and each revolves around a stator or rotor slot with counter-clockwise as positive. Any stator/rotor pole pair is connected with reluctance element R_{ag}, and the number of airgap connections equals to numerical sum of stator and rotor teeth $(N_s + N_r)$.

The feature of the MEC model in Fig. 2.43 is that the airgap reluctance R_{ag} connects a couple of rotor teeth from one stator tooth. Identification of R_{ag}, the reverse of permeance P_{ag}, on each connecting line is a key step, for which a novel definition of rotor position is illustrated. In Fig. 2.44, geometrical relation of a pole pair is defined. The misalignment x is defined as the horizontal distance between centerlines c_1 and c_2 of the pole pair. The maximum x occurs when teeth are aligned while the minimum is when adjacent rotor tooth is aligned with the pairing stator tooth. The x value is deducted where specifically τ_s and τ_r are respectively stator and rotor pole pitch. The flux through airgap Φ is regarded going via the straight lines that connect nodal of each pole pair. Instead of using conventional definition of aligned and unaligned position as positional boundary of a pole pair, the range is redefined from the total alignment to the misalignment position that the stator tooth of that pair is aligned with the next rotor tooth, the reason that any of the stator poles will get connected with a couple of adjacent rotor poles. The range is quantitated by misalignment x value that can be obtained when stator and rotor pole pitch are known.

R_{ag} is obtained from $R = \Theta/\Phi$. The airgap flux Φ is defined as $\Phi = L_z(\varphi_{l1} - \varphi_{l2})$, where L_z is the axial length of the machine, φ_{l1} and φ_{l2} are magnetic vector potentials at l_1 and l_2 (Fig. 2.43). The MMF Θ between each pole pair connection is obtained by integrating radial component of magnetic field along the straight lines (i.e., AB in Fig. 2.43) via FE method. The FE is performed using FEMM program controlled via Matlab console. In simulation, a tooth pair moves with incremental steps in the full range of x under constant phase current. The integration along the straight lines is easily done by creating that line that connects nodal points of a pole pair.

R_{ag} serves as a function of misalignment x. To make $R_{ag}(x)$ dimensionless, relations are defined as follows, $x^* = x/\tau_{av}$ where $\tau_{av} = (\tau_s + \tau_r)/2$ is the average pole pitch, and $R^*_{ag}(x) = R_{ag}(x)/(\mu_0\tau_{av}/\delta)$ where δ is the airgap length and μ_0 is permeability of vacuum. As a result, analytical function $R^*_{ag}(x^*)$ is given on the studied motor to make the fitting function suited to geometrical variation. It is found that $R^*_{ag}(x^*)$ can be fitted by a polynomial of 5 degrees of freedom (d_1 to d_5). $\Sigma d_i = 1$, where $i = 1$–6. $R^*_{ag}(x^*)$ is a four-order polynomial [42]. The calculation is linear without magnetic saturation, meaning that the magnetic curves are a group of straight lines, starting from zero. Hence, the developed fitting approach is able to estimate magnetic reluctance without FE for an arbitrary machine geometry.

The magnetic saturation is calculated by a fitting approach in Sect. 2.4, with the same lookup in Fig. 2.36. The studied 3-phase 12/8 SRM is from Table 2.9. Phase inductance variation is calculated and shown in in Fig. 2.45 and FE is used as reference. The tendency decreases from the aligned (0°) to unaligned rotor position (22.5°). Through computational deviation at some points occurs, the overall variation is acceptable. In particular, both methods agree when phase current I_{ph} is very low, i.e. $I_{ph} = 0$, or 45A. In such an occasion, the magnetic saturation is quite unlikely to happen regardless of rotor position, indicating that the proposed curve fitting of the airgap permeance is quite accurate. However with current increasing, the numerical difference occurs and the gap is enlarged with higher excitation, especially when teeth are partially overlapping, i.e. from $I_{ph} = 90$ till 180 at 5°. Such phenomenon lies that magnetic saturation tends to be subjected to uneven distribution on poles. When the distribution is more uniform, i.e. when poles are aligned (close to 0°), the gap in value reduces as well.

The fitted curve and surface together illustrate that the saliency effect that contributes to complex airgap flux and distributed magnetic saturation can be modeled in a simple approach. To verify the accuracy of fitting, flux linkage curves

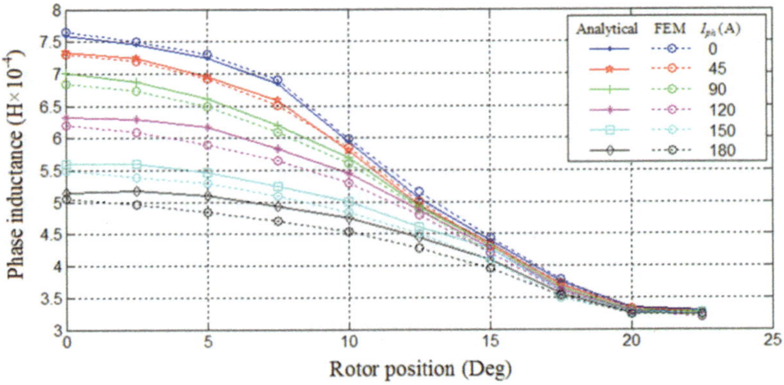

Fig. 2.45 The phase inductance-position relationship by the proposed fitting model

Table 2.10 The machine geometry variation

	β_s	β_r	δ
M0	20°	20°	0.5 mm
M1	20°	30°	0.5 mm
M2	20°	20°	0.25 mm

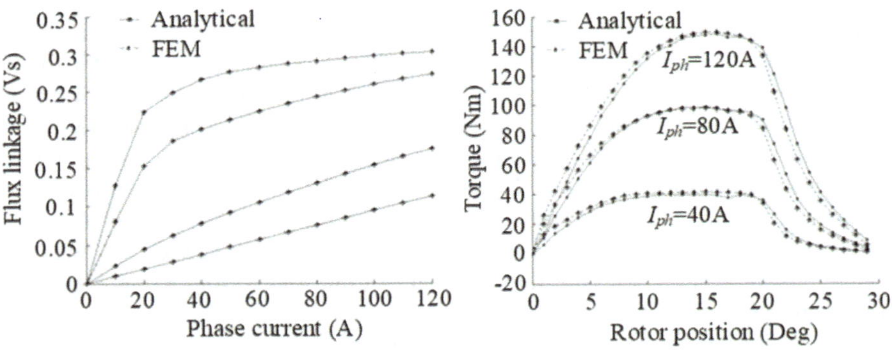

Fig. 2.46 The magnetic curves and phase torque of the machine M0 in Table 2.10

are calculated. The curves are incremented by 1° rotation, indicating that accuracy under any given operation condition is required very high. To prove the flexibility of the algorithm, the number of turns per pole is 35 and the maximum phase current is 120A, making magnetic saturation sufficiently high. This machine serves as M0 and principal geometrical variation is shown in Table 2.10. Constant one-phase current at any rotor position is applied.

Typical machines M0–M2 in Table 2.10 are analyzed. Such machines are featured by variation of airgap and widths of poles, the most influential factors. The circuit loop permeance is calculated based on the airgap fitting subtracted by the drop of magnetic saturation P_d. The magnetic curves are shown in Figs.2.46, 2.47 and

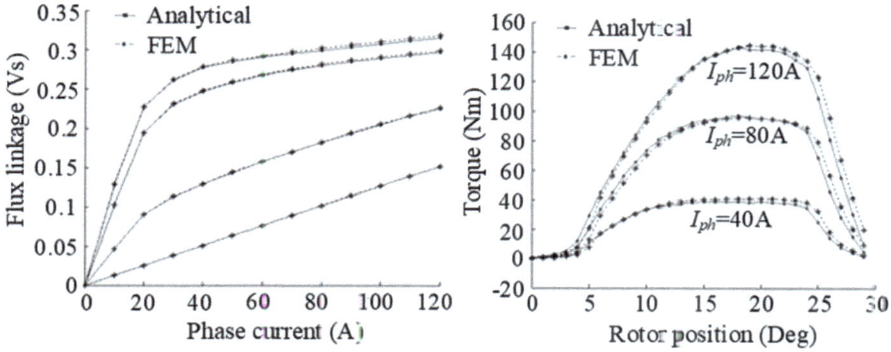

Fig. 2.47 The magnetic curves and phase torque of the machine M1 in Table 2.10

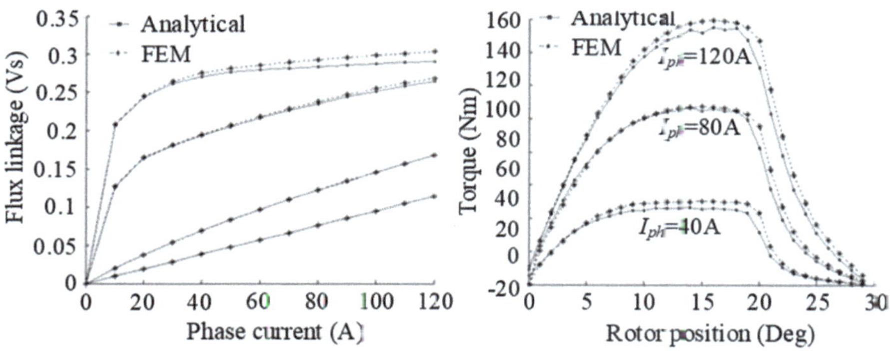

Fig. 2.48 The magnetic curves and phase torque of the machine M2 in Table 2.10

2.48 respectively. Starting from aligned position 0°, 1° increment of rotor position till the unaligned 30° is calculated. To facilitate comparison, only 4 typical curves representing the aligned, unaligned, partial overlap and before start of overlap, are presented. It is shown agreement between FE and proposed method. Further, phase torque is calculated from magnetic curves with energy conversion principle. The torque curves at different phase current levels are further obtained based on the change of flux linkage under very small incremental steps of rotation. The torque curves are quite smooth without ripples, showing high accuracy of the fitting method.

2.6 Chapter Summary

Fast and accurate calculation of saliency effect of switched reluctance machines are mathematically challenging. The flux linkage calculation is required very accurate. For SRMs, the curves with small incremental of rotor position and phase current should be accurately identified. The traditional magnetic equivalent circuit model is a classical solution which however is not accurate on SRM due to the double saliency structure that leads to rotor position dependent airgap reluctance and distributed magnetic saturation. In this chapter, improvements are made based on the network circuit model. In Sect. 2.2, airgap flux is categorized into up to 7 typical paths while salient poles are accordingly discretized, and further can shields are added as update. Desired calculation accuracy on flux linkage is obtained for an ordinary and then canned machine. In Sect. 2.3, this model is further updated by variable loss coefficients and iron loss is calculated. In Sect. 2.4, the modeling of magnetic saturation distribution on poles is simplified using a fitting method, which also achieves high calculation. Further in Sect. 2.5, it is found that the airgap permeance can be also fitted, using a four-order polynomial.

By applying improved network models, flux linkage characteristics can be calculated in a simple and accurate way, especially at intermediate rotor positions. Computation time is reduced significantly and high accuracy still remains. Such method also applies to magnetic devices of salient poles, especially with high flux density.

References

1. C.C. Chan, The state of the art of electric and hybrid vehicles. Proc. IEEE **90**(2), 247–275 (2002)
2. Z.Q. Zhu, D. Howe, Electrical machines and drives for electric, hybrid and fuel cell vehicles. Proc. IEEE **95**(4), 745–765 (2007)
3. N. Schofield, S.A. Long, D. Howe, M. McClelland, Design of a switched reluctance machine for extended speed operation. IEEE Trans. Ind. Appl. **45**(1), 116–122 (2009)
4. A. Radun, Analytical calculation the switched reluctance motor's unaligned inductances. IEEE Trans. Magn. **35**(6), 4473–4481 (2001)
5. A. Radun, Analytically computing the flux linked by a switched reluctance motor phase when the stator and rotor poles overlap. IEEE Trans. Magn. **36**(4), 1996–2003 (2002)
6. D.A. Torrey, Modeling a nonlinear variable-reluctance motor drive. Proc. IEEE **137**(5), 314–326 (1990)
7. P. Rafajdus, I. Zrak, V. Hrabovcova, Analysis of switched reluctance motor parameters. J. Electr. Eng. **55**(7), 195–200 (2004)
8. R. Krishman, *Switched Reluctance Motor Drives* (CRC Press, Boca Raton, 2001), pp. 30–77 ISBN:0-8493-0838-0
9. G. Dajaku, D. Gerling, Stator slotting effect on magnetic field distribution of salient pole synchronous permanent machines. IEEE Trans. Magn. **46**(9), 3676–3683 (2013)
10. U. Kim, D.K. Lieu, Magnetic field calculation in permanent magnet motors with rotor eccentricity and with slotting effect. IEEE Trans. Magn. **34**(4), 2253–2266 (1998)
11. S. Sudhoff, B. Kuhn, K. Corzine, B. Branecky, Magnetic equivalent circuit modeling of induction motors. IEEE Trans. Energy Converse. **22**(2), 259–270 (2007)
12. B. Asghari, V. Dinavahi, Experimental validation of a geometrical nonlinear permeance network based real-time induction machine model. IEEE Trans. Ind. Electron. **59**(11), 4049–4062 (2012)
13. C.S. Edrington, B. Fahimi, M. Krishnamurthy, An auto calibrating inductance model for switched reluctance motor drives. IEEE Trans. Ind. Electron. **54**(4), 2165–2173 (2007)

14. J. Corda, S.M. Jamil, Measurements of equivalent circuit parameters for modeling the impact of iron losses in SR machine, International Conference on Electrical Machines, ICEM 2003. Vilamoura, Portugal
15. C.P. Weiss, M. Huebner, M.D. Henner, W.D. Doncker, Switched reluctance machine model considering asymmetries and enabling dynamic fault simulation, IEEE International Conferences on Electrical Machines & Drives ICEMS 2013, pp. 979–985
16. G. Stimberger, B. Polajzer, D. Dolinar, Evaluation of experimental methods for determining the magnetically nonlinear characteristics of electromagnetic devices. IEEE Trans. Magn. **41**(10), 4030–4032 (2005)
17. C. Cossar, M. Popescu, T.J.E. Miller, M. McGilp, On-line phase measurements in switched reluctance motor drives, Proceedings of EPE 2007, Aalborg, Denmark
18. K. Atallah, D. Howe, Calculation of rotational power loss in electrical steel laminations from measured HB. IEEE Trans. Magn. **29**(6), 3547–3549 (1993)
19. T. Kochmann, Relationship between rotational and alternating losses in electrical steel sheets. J. Magn. Magn. Mater. **160**(1), 145–146 (1996)
20. D. Gerling, Comparison of different methods for calculating airgap field of inductance motors. J. Electr. Eng. **77**(2), 101–106 (1994). Springer
21. G. Dajaku, Electromagnetic stator slotting effect of salient pole synchronous permanent magnet machine. IEEE Trans. Magn. **44**(5), 1528–1537 (2012)
22. N. Sadowski, M.L. Mazenc, Evaluation and analysis of iron losses in electrical machines using the rain-flow method. IEEE Trans. Magn. **36**(4), 1923–1926 (2000)
23. D.M. Ionel, M. Popescu, S.J. Dellinger, Computation of core loss in electrical machines using improved models for laminated steel. IEEE Trans. Ind. Appl. **43**(6), 1554–1564 (2007)
24. H. Domeki, Y. Ishihara, C. Kaido, Y. Kawase, Investigation of benchmark model for estimating iron loss in rotating machines. IEEE Trans. Magn. **40**(2), 794–797 (2004)
25. Y. Hayashi, T.J.E. Miller, A new approach to calculate core losses in the SRM. IEEE Trans. Ind. Appl. **31**(5), 1039–1046 (1995)
26. Z. Gmyrek, A. Boglietti, A. Cavagnino, Iron loss prediction with PWM supply using low and high frequency measurements: analysis and results comparison. IEEE Trans. Ind. Electr. **55**(4), 1722–1728 (2008)
27. Z. Gmyrek, A. Boglietti, A. Cavagnino, Estimation of iron losses in induction motors: calculation method, results, and analysis. IEEE Trans. Ind. Electron. **57**(1), 161–171 (2010)
28. Z.Q. Zhu, L.J. Wu, Z.P. Xia, An accurate subdomain model for magnetic field computation in slot surface mounted permanent-magnet machines. IEEE Trans. Magn. **46**(4), 1110–1115 (2010)
29. T. Lubin, S. Mezani, A. Rezzoug, Exact analytical method for magnetic field computation in the air gap of cylindrical electrical machines considering slotting effects. IEEE Trans. Magn. **46**(4), 1092–1099 (2010)
30. D. Gerling, G. Dajaku, Three-dimensional analytical calculation of induction machines with multilayer rotor structure in cylindrical coordinates. Electr. Eng. **86**(4), 199–211 (2004). Archiv fuer Elektrotechnik, Springer
31. A. Boglietti, A. Cavagnino, M. Lazzari, M. Pastorelli, Predicting iron losses in soft magnetic materials with arbitrary voltage supply: an engineering approach. IEEE Trans. Magn. **39**(2), 981–989 (2003)
32. M.J. Manyage, P. Pillay, Low voltage high current PM traction motor design using recent core loss results, IEEE Industry Application Annual Meeting, 2007, pp. 1560–1566
33. K. Atallah, Z.Q. Zhu, D. Howe, An improved method for predicting iron losses in brushless permanent magnet DC drives. IEEE Trans. Magn. **28**(5), 2997–2998 (1992)
34. P. Materu, R. Krishnan, Estimation of switched reluctance motor losses. IEEE Trans. Ind. Appl. **28**(3), 79–90 (1992)
35. G. Bertotti, A. Canove, M. Chiampi, D. Chiarabaglio, F. Fiorillo, A.M. Rietto, Core loss prediction combining physical models with numerical field analysis. J. Magn. Mater. **133**(1), 647–650 (1994)

36. S. Steentjes, M. Lebmann, K. Hameyer, Advanced iron-loss calculation as a basis for efficiency improvement of electrical machines in automotive application. IEEE Trans. Ind. Appl. **41**(4), 1–6 (2012)
37. J.D. Lavers, P.P. Biringer, H. Hollitscher, A simple method of estimating the mirror loop hysteresis loss in thin laminations. IEEE Trans. Magn. **14**(5), 386–388 (1978)
38. ANSYS Corporation, Maxwell user's guide (2017), http://support.ansys.com/
39. Q. Yu, C. Laudensack, D. Gerling, An analytical network for switched reluctance machines with highly saturated regions, IEEE Electrical Machines and Power Electronics and Electromotion Joint Conference, ACEMP 2011, Istanbul Turkey
40. J.M. Kokernak, D.A. Torrey, Magnetic circuit model for the mutually coupled switched reluctance machine. IEEE Trans. Magn. **36**(2), 302–309 (2000)
41. J.F. Farshad, C. Lucas, Development of analytical models of switched reluctance motor in two-phase excitation mode: extended miller model. IEEE Trans. Magn. **41**(6), 2145–2155 (2005)
42. Q. Yu, X.S. Wang, Y.H. Cheng, Determination of air-gap flux density characteristics of switched reluctance machines with conductor layout and slotting effect. IEEE Trans. Magn. **52**(8), 1 (2016)

Chapter 3
Electromagnetic Analysis of Can Effect of a Canned SRM

3.1 Canned Switched Reluctance Machine and Operation Principles

In this chapter, electromagnetic analysis of a canned SRM is studied. Finite element (FE) method is applied, with validation via measurement. Chapter organization is as follows: In Sect. 3.1, canned SRM and operation principle are briefly introduced with the calculation method described. In Sect. 3.2, eddy current distribution at defined typical rotor positions by a single phase excitation is illustrated, followed by applying all phases. In Sect. 3.3, the can loss analysis is extended to one stroke period and further different control modes. In Sect. 3.4, the airgap flux and eddy current loss features due to the use of cans are shown. In Sect. 3.5, experimental validation is taken.

Figure 3.1 shows a general view of a 3-phase 12/8 canned SRM, with airgap radial length enlarged disproportionately. The machine geometry stems from Table 2.8. Compared with a standard design, airgap is enlarged to accommodate cans. There are a couple of cans, namely the stator can and the rotor can. The stator can is shrunk into the stator bore while the rotor can is clamped onto rotor poles and rotates synchronously. When the cooling liquid goes between cans, the stator can is to protect armature coils while rotor can is to reduce inertia and friction. Each can shield is 0.3 mm in radial length approximately, the minimum value of manufacturing, to reduce the airgap length and ohmic can loss while maintaining mechanical stiffness. The material is Hastelloy C, a nickel-based non-magnetic alloy with high electrical resistivity. The value is 1.25 μΩm at 20 °C and is approximately 1.4 times higher at every 100 °C rise. Hence, the total airgap including cans is conductive but non-magnetic, and thus only eddy current loss, instead of hysteresis loss occurs.

The energy conversion principle now is featured by the use of cans. The cans serve as additional energy consumption parts, reducing magnetic flux density. Equations (3.1) and (3.2) show flux linkage ψ when a phase is respectively fluxing

Q. Yu et al., *Analysis and Mathematical Models of Canned Electrical Machine Drives*, https://doi.org/10.1007/978-981-13-2745-2_3

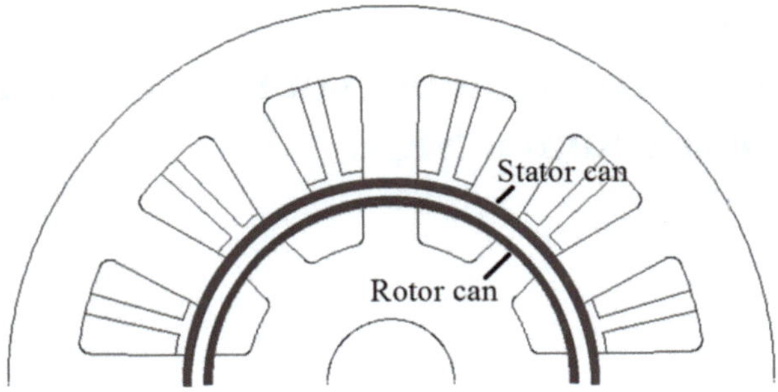

Fig. 3.1 Sketch of the studied canned SRM

and defluxing, with voltage drops by resistances from armature coils (r_a) and in particular by cans including the stator can r_{sc} and rotor can r_{rc}. The energy equations are,

$$\psi = \int_{\theta_{on}}^{\theta_{off}} (V_s - r_a i - \varphi)dt = \int_{\theta_{on}}^{\theta_{off}} (V_s - r_a i - r_{sc} i_{sc} - r_{rc} i_{rc})dt \quad (3.1)$$

$$\psi = \int_{\theta_{off}}^{\theta_{ext}} (-V_s - r_a i + \varphi)dt = \int_{\theta_{off}}^{\theta_{ext}} (-V_s - r_a i + r_{sc} i_{sc} + r_{rc} i_{rc})dt \quad (3.2)$$

where θ_{on}, θ_{off} and θ_{ext} are rotor positions corresponding to phase turn-on, turn-off and current extinguishment respectively. V_s is voltage supply. The time dependent i, i_{sc} and i_{rc} refer to armature phase current, induced eddy current on the stator can and rotor can respectively. r_{sc} and r_{rc} are electrical resistance of a circulation path for the induced eddy current on cans and are time dependent as well. The current and path are presented by one additional flux φ, indicating the use of cans that weakens the main flux linkage ψ.

An inherent feature of SRM is that the phase current works at intervals divided by rotor pole pitch. If voltage drops in (3.1) and (3.2) including cans are taken, the dwell period is increased, making control strategy of a canned SRM different. Figure 3.2 shows phase current waveforms with the rated current 60A, respectively by fixed rotor position control at low speed, turning-on advancing at mediate speed and continuous excitation at high speed. Generally, the time period to deflux the machine should equal to flux. In Fig. 3.2a, current begins before start of the pole pair overlap and terminates before alignment. It is inappropriate to further extend that dwell angle, as subsequent residual flux leads to reverse torque. With speed increase, the rate of change of phase current i/θ is reduced (Fig. 3.2b). Hence, the turn-on advancing control is used to magnetize in advance while to reduce residual flux

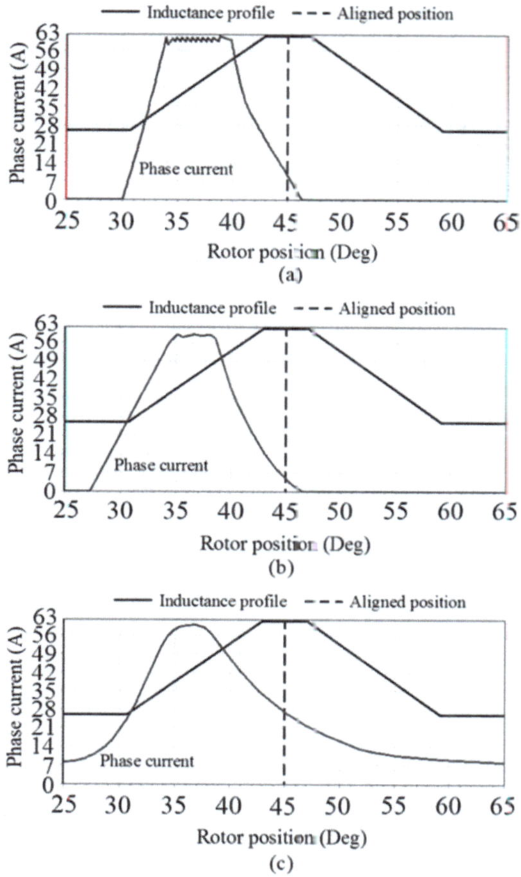

Fig. 3.2 Phase current with rated hysteresis band 60A, (**a**) fixed control mode at speed 2000 rpm, (**b**) turn-on advancing control mode at 3000 rpm, (**c**) continuous excitation control mode at 4500 rpm

after alignment. Under higher speed, phase current will continue across subsequent phase, leading to phase being fluxed all the time during commutation (Fig. 3.2c).

The eddy current is analyzed by FE method. The airgap and cans are sensitive components that need fine mesh for higher calculation accuracy. In the FE model, there are 3 concentric cylindrical thin layers for the stator can, airgap and rotor can respectively. Each layer is circumferentially divided into 240 pieces to confine dimensions of meshed elements. The can loss is calculated by a direct discretization method. Due that the magnetic flux variation is subjected to difference at each location, the machine is discretized into i elements with flux density $B_i(t)$. Further for each element, flux variation is not sinusoidal and conventional Steinmetz model does not apply. For i-th element, the loss density E_i is directly seen as a function of square rate of change of flux density $B_i(t)$, which is

$$E_i = \frac{1}{T}\int_0^T \frac{c}{2\pi^2}\cdot\left(\frac{dB_i(t)}{dt}\right)^2 \cdot dt \tag{3.3}$$

where c refers to the Steinmetz coefficient and T is the time period. Further, Eq. (3.3) is discretized as

$$E_i = \frac{c}{2\pi^2}\cdot\frac{1}{T\cdot\Delta t}\sum_{k=1}^{N}(B_{i\_k+1} - B_{i\_k})^2 \tag{3.4}$$

where N is the number of time steps per time period T, and B_i is the flux density of i-th element, Δt is the time interval between the discretized k and $k+1$ steps, and it is defined as

$$\Delta t = t_{k+1} - t_k = \frac{T}{N} = \frac{1}{f\cdot N}, \quad k = 1, 2, \ldots\ldots, N \tag{3.5}$$

where f is the frequency. Combining (3.4) and (3.5), the loss density of i-th element is written as

$$E_i = \frac{c}{2\pi^2}Nf^2\sum_{k=1}^{N}(B_{k+1\_i} - B_{k\_i})^2 \tag{3.6}$$

Then, the loss P is written as

$$P = \sum_{i=1}^{M} m_i E_i = \frac{c}{2\pi^2}Nf^2\sum_{i=1}^{M}\left(m_i\sum_{k=1}^{N}(B_{k+1\_i} - B_{k\_i})^2\right) \tag{3.7}$$

where m_i is the mass of i-th element, M is the total number of elements, B_{k_i} and B_{k+1_i} are flux density of i-th element B_i at time step k and $k+1$ respectively. With flux density in 3 dimensions, the eddy current loss is written as

$$P = \frac{c}{2\pi^2}Nf^2\sum_{i=1}^{M} m_i \begin{pmatrix} +\left(\sum_{k=1}^{N}(B_{x\_k+1\_i} - B_{x\_k\_i})^2\right) \\ +\left(\sum_{k=1}^{N}(B_{y\_k+1\_i} - B_{y\_k\_i})^2\right) \\ +\left(\sum_{k=1}^{N}(B_{z\_k+1\_i} - B_{z\_k\_i})^2\right) \end{pmatrix} \tag{3.8}$$

where $B_{x\_k\_i}$ and $B_{x\_k+1\_i}$, $B_{y\_k\_i}$ and $B_{y\_k+1\_i}$ and $B_{z\_k\_i}$ and $B_{z\_k+1\_i}$ are flux density at corresponding step for x-y-z direction respectively.

3.2 Eddy Current and Loss Features at Typical Rotor Positions

To identify eddy current and loss due to the use of cans, a couple of typical rotor positions is defined, one close to the unaligned position and the other at partial aligned position (Fig. 3.3). Note that the curvature is neglected and rotor is moving leftward. According to operation principle, Coil 1 is fluxing while Coil 2 is defluxing, such that fast variation of airgap flux occurs. Such flux goes through cans, causing main eddy current and loss. In other words, magnetic phase fluxing and defluxing at this couple of positions are typical moments. In this section, electromagnetic analysis is taken at typical positions. Single phase excitation is studied, followed by all phases.

3.2.1 *Single Phase Excitation*

The can shield with eddy current distribution at typical rotor positions is shown in Fig. 3.4. Due that cans are quite close and imposed in similar magnetic flux field, only the stator can is shown. Due to symmetry, only half of the can shield along axial direction is shown. High eddy current circulation occurs in each case, which is around "stator pole" region, the joint of the can and the fluxing/defluxing stator pole. Note that: (1) there are 4 symmetrical circulations in parallel, as totally 4 pole pairs of a 12/8 SRM are excited at this moment; (2) For each circulation, the highest density occurs at the axial end, causing end effect; (3) Distribution on upper and lower side of the "stator pole" region is unnecessarily symmetry, because the excited stator and rotor poles are not fully aligned, which affects airgap flux path. Comparatively, eddy current density when defluxing (Fig. 3.4b) is much higher than fluxing (Fig. 3.4a), as airgap resistance differs sharply with rotor positions. During defluxing, the pole pair is partially aligned and the resistance is much lower, causing a faster change of airgap flux and higher current.

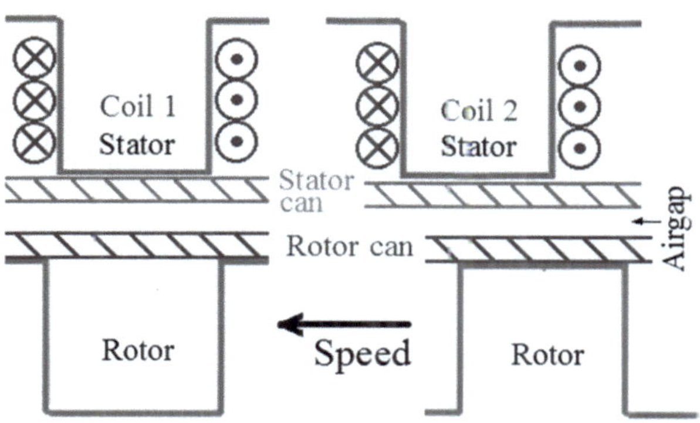

Fig. 3.3 Definition of typical rotor positions of phase fluxing and defluxing

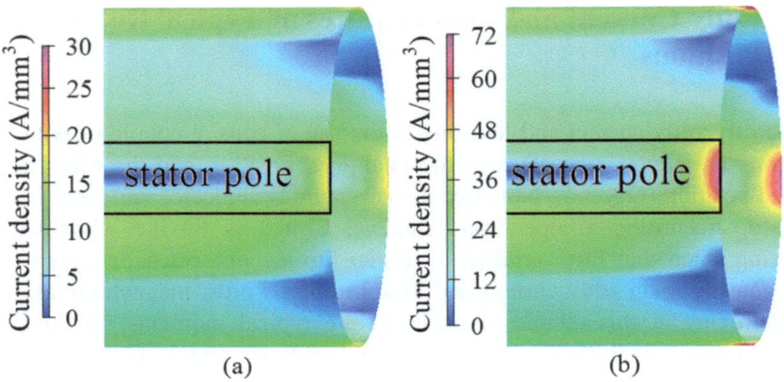

Fig. 3.4 Eddy current circulation on stator can by single phase excitation, where phase current corresponds to Fig. 3.2a; (**a**) The secondary circulation at the moment of fluxing; (**b**) The principal circulation at the moment of defluxing. The rectangular zone indicates the joint of the can and the flux/defluxing stator pole. Note that the numerical scale in (**b**) is more than twice of (**a**)

3.2.2 All-Phase Excitation

Eddy current distribution on both rotor/stator core and the can shields of the studied machine is analyzed under all-phase excitations. Due to extremely high loss on the cans, it is impossible to satisfactorily show the distribution of both parts in one figure and therefore it must be separately shown. Each phase is in sequence fluxed and defluxed and in analogy some typical rotor positions should be identified first. According to operation principle, there exist one moment where magnetic fluxing and defluxing from different phases simultaneously occur. That is, there exists one highly typical rotor position that includes both cases in Fig. 3.4. Based on Section A, this position is analyzed by overlapping the couple of single phase excitations.

Figure 3.5 shows the machine, with eddy current loss density distributed on rotor/stator at this typical position. Peak density locates on the partially aligned defluxing pole pair, due to the fast change of airgap flux with a small airgap. Meanwhile the adjacent almost unaligned pole pair has the secondarily high loss density due to phase fluxing with a large airgap.

Figure 3.6 further shows eddy current circulation on the stator can. The view corresponds to 90° clockwise rotation along Y-axis from Fig. 3.5. Principal and secondary eddy current circulations in Fig. 3.4b and Fig. 3.4a are simultaneously imposed. As there is a 30° angular displacement between the fluxing and defluxing stator pole, this couple of circulations is partially overlapped with this displacement. In Fig. 3.6, the principal circulation is in analogy around "stator pole" region, whereas the secondary is not clearly seen, due to the sharply reduced numerical values. Such overlapped circulations, one by phase fluxing and the other by defluxing, are in reverse revolution directions, indicating that phases are dependent and eddy current at some locations may be reduced by offset. Such a feature offers a

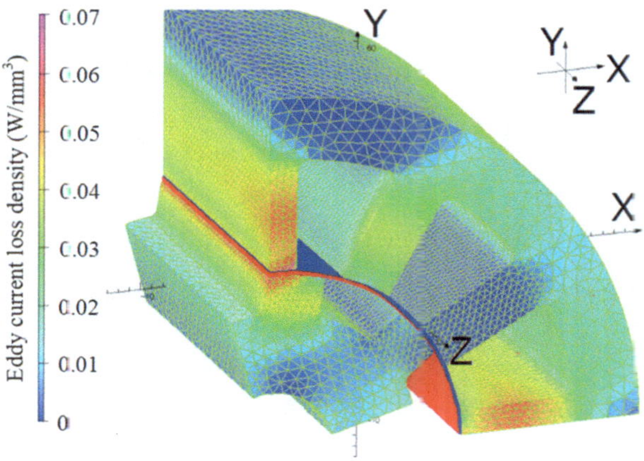

Fig. 3.5 Geometric overview of the canned machine and distribution of eddy current loss density on rotor/stator, showing typical loss generation due to phase fluxing and defluxing simultaneously; Note the definition of X-Y-Z coordinate, where Z is the axial direction

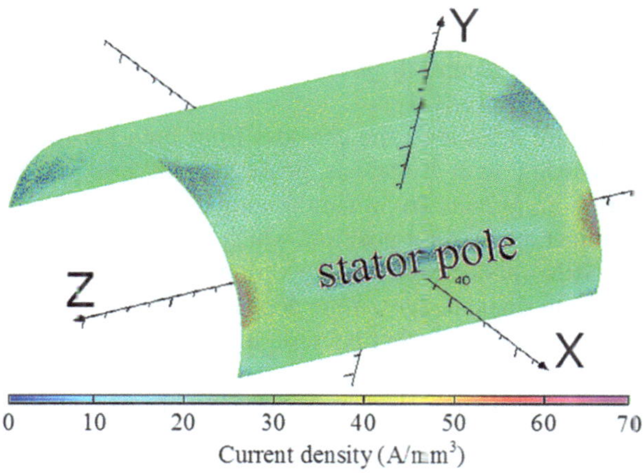

Fig. 3.6 The eddy current distribution on the can shield, at the rotor position corresponding to Fig. 3.5, showing the overlapped current circulations by both phase fluxing and defluxing; Note that the spatial position of the X-Y-Z coordinate corresponds to 90° clockwise rotation along Y-axis from the coordinate in Fig. 3.5

unique loss reduction method. That is to redefine control strategy regarding turn-on and turn-off angles to optimize rotor position where circulations occur.

3.3 Can Loss Variation of One Stroke Period

In this section, can loss based on eddy current circulation is investigated, which is not limited to the couple of typical rotor positions but extended to full range. The range is identified by one stroke, which refers to one period of phase excitation and is 15° in mechanical degrees for a 3-phase 12/8 SRM. The rotor position 22.5° is defined as the unaligned while 45° is aligned position. According to the can loss variation, the range 34°–49° is selected as one stroke. Low speed operation at fixed control mode corresponding to Fig. 3.2a is first studied, followed by the medium and

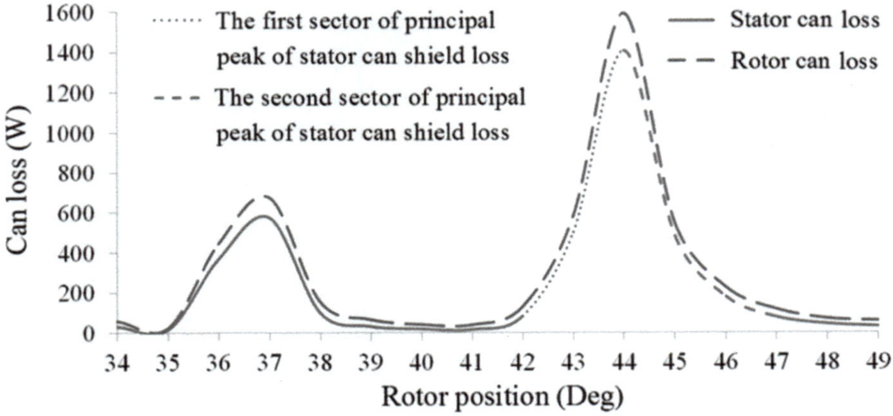

Fig. 3.7 The can loss variation at low speed

high speed regarding phase advancing in Fig. 3.2b and continuous excitation in Fig. 3.2c respectively.

Figure 3.7 shows the can loss from both the rotor and stator can at low speed 2000 rpm, with hysteresis phase current level 60A. Both curves share similar variation, as cans are in similar airgap flux field. Note that loss from the rotor can is higher at any rotor position, which is more obvious at the peak. This is because of the SRM saliencies that make variation of airgap flux close to rotor pole, where the rotor can shield resides, more dependent on rotor positions and hence a faster change of flux and more loss. Likewise, the stator can is analyzed in the following.

In Fig. 3.7, the curve is featured by a couple of peaks, forming the principal part of loss. In analogy, peaks are donated to the principal and secondary. The principal within 42°–46° stems from phase defluxing while the secondary within 35°–38° from phase fluxing. Both peaks are due to the fast change of airgap flux when phases are undergoing the current change by turn-on or turn-off. The loss of the secondary is comparatively lower, due to the unaligned pole pair that brings high airgap reluctance that helps to resist the rate of change of flux. In the following, the principal peak by phase defluxing is discussed.

The principal peak is further divided into a couple of sectors, according to whether loss generation stems from the on-going phase defluxing or residual flux thereafter. The first sector starts at 42° when the defluxing starts, rises up sharply, and terminates at 44° just before alignment where defluxing ends. Figure 3.8a shows accordingly the detailed eddy current distribution on the stator can. Similar to Fig. 3.6, the eddy current revolves around the defluxing phase, with high density at the end part. The second sector starts immediately at 44° and ends till 46°. During this sector, defluxing doesn't exist while eddy current on the can perpetuates. Accordingly the variation takes on a smooth instead of an abrupt decline. The perpetuation stems from the electromagnetic coupling that causes residual flux and

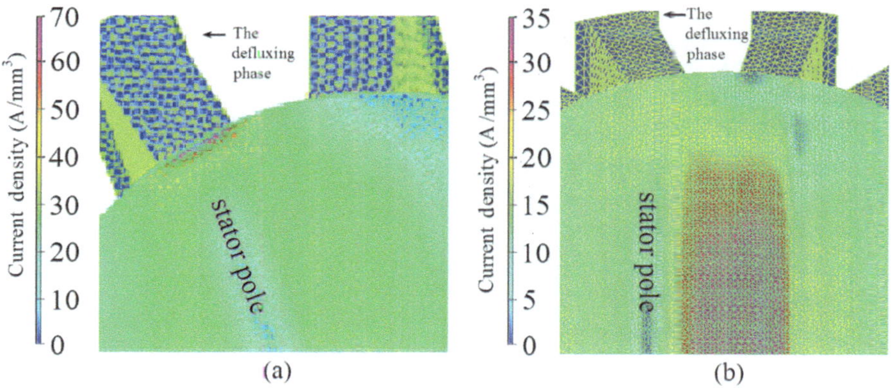

Fig. 3.8 Eddy current distribution on the stator can when the can loss is undergoing the principal peak, with stator poles and in particular the defluxing one shown as reference, (**a**) at the first sector with phase defluxing; (**b**) at the second sector with residual flux; The phrase "stator pole" means the joint zone of the stator can and the defluxing stator pole. Note that the numerical scale in (**a**) is twice of (**b**)

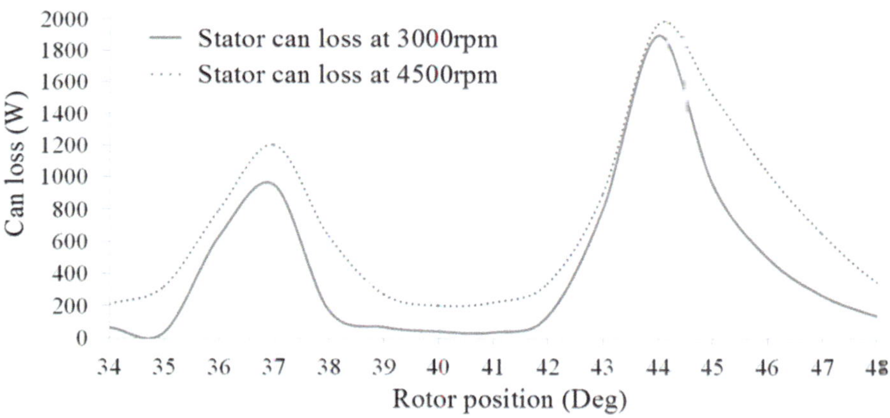

Fig. 3.9 Loss variation of the stator can at medium and high speed

eddy current. Figure 3.8b shows the current distribution and note that the numerical values are lower compared to Fig. 3.8a.

With rotor speed increasing, phase advancing control mode is applied, followed by continuous excitation across phases. Figure 3.9 shows loss variation of the stator can at medium speed 3000 rpm and high speed 4500 rpm, with the same hysteresis current level 60A. Likewise, both cases are featured by a couple of peaks. In comparison, the loss at any rotor position is higher under higher speed, due to a faster change of airgap flux.

At high speed, the curve takes on characteristic features. First, for the second sector of the principal peak, the rate of decline is reversely reduced. This is due to

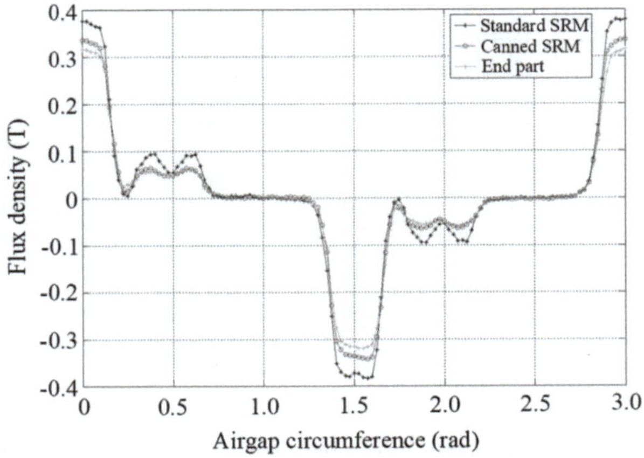

Fig. 3.10 Airgap flux distribution of a standard and a canned SRM at the typical rotor position corresponding to Fig. 3.5

intensified residual eddy current on cans that alleviates the rate of change of airgap flux. Next, a sharp increase of loss at positions out of the peaks such as 34°–35° and 38°–42° occurs, due to residual flux across phases by the continuous excitation mode.

3.4 Airgap Flux and Eddy Current Loss Due to the Use of Cans

Airgap flux serves as the central consideration in magnetic analysis, and it is affected by saliency with magnetic saturation. When an additional can is inserted, flux is further affected by the induced eddy current on cans. In this section, the change of airgap flux features due to the use of cans is studied, followed by consequent eddy current loss in rotor/stator core. To this end, a couple of SRMs, except one of them is canned, is studied.

Figure 3.10 shows airgap flux density distribution of a standard and a canned SRM, with the same phase current and at the typical rotor position corresponding to Fig. 3.5, where phase fluxing and defluxing simultaneously occur. Likewise, each waveform is featured by a couple of peaks, the principal and secondary. The principal is excited by the defluxing pole pair while the secondary peak is by the fluxing pair. Comparatively there are three features. (1) For the standard SRM, note that there exists a cavity on the secondary peak, indicating that at unaligned position most flux lines go along each side, instead of middle axis line, of a stator pole. When the machine is canned, the cavity is reduced, as the induction on the can will reduce flux harmonics. (2) An overall decrease for the canned machine occurs, in particular at both peaks. This is because of the can induction that weakens airgap flux. (3) As to the flux waveform at the end part, numerical values are further reduced. This is

Table 3.1 Comparison of eddy current loss on stator and rotor core of a standard and a canned SRM, *S* standard, *C* Canned

Speed (krpm)	Loss (W) with temperature		
	S. SRM 20°C	C. SRM 20°C	C. SRM 90°C
0.8	32.2	30.4	29.5
1.6	65.3	60.1	59.1
2.4	112.2	104.0	102.5
3.2	170.7	158.3	156.5
4.0	244.5	226.8	224.6

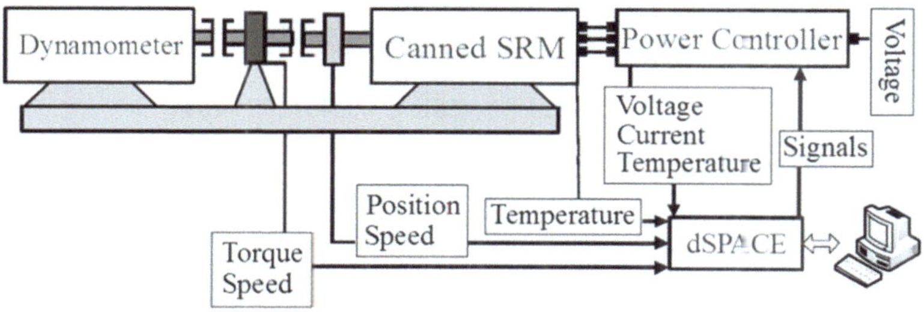

Fig. 3.11 The testing system setup

because of the intensified induction at the end part of the cans that weakens airgap flux.

The airgap flux variation due to the use of cans will lead to a change of eddy current loss in rotor/stator core. Such loss is calculated from (3.5), (3.6) and (3.7) and is shown in Table 3.1. With the same temperature 20 °C, both cases share similarity at each speed, which means the use of cans has limited electromagnetic effect on eddy current loss generation. Note that loss of the canned machine is lower, due to the reduced flux density and less flux harmonics. Table 3.1 shows that when temperature jumps to 90 °C (cooing liquid), loss is hardly reduced, which means the use of cans has limited thermal effect on eddy current loss as well.

3.5 Experimental Validation

The can loss generation is verified by experiment and the system structure is shown in Fig. 3.11. A 20 kW 12/8 SRM is connected to a DC load dynamometer for torque measurement. The hardware dSPACE 1103 with FPGA and a piggy board commands SRM power controller based on voltage, current and temperature feedback signals as well as control from a computer. The FPGA port provides a clock rate up

Table 3.2 Friction loss by coast down test, *S* standard, *C* canned

Speed (krpm)	Loss (W)	
	S. SRM	C. SRM
0.6	1.3	14.1
1.2	3.0	33.2
1.8	4.8	55.2
2.4	7.1	89.0
3.0	9.6	132.5
3.6	12.7	197.7

Table 3.3 Copper loss due to the use of cans, *S* standard, *C* canned

Percent (%)	Percent (%)			
	1krpm	1krpm	3krpm	3krpm
	S. SRM	C. SRM	S. SRM	C. SRM
40.4	89	92	87	93
70.3	252	261	240	264
99.6	501	523	483	572

to 100 MHz, allowing real-time control of power converter. Control modes include fixed rotor position, phase advancing and continuous excitation, and are carried out via Matlab/Simulink. For sake of comparison, another standard machine of the same geometry, but without cans is built.

For experiment setup, the machine is running under the worst case of no liquid cooling between cans. Another feature is that the airgap friction loss should be taken, due to close distance and high relative movement between cans. The friction loss is from a coast down test [1]. The value is seen time averaged, which is determined by rotor inertia, shaft speed and retarding time. Table 3.2 shows the friction loss rises up geometrically with rotor speed. Loss of the canned machine is comparatively much higher due to the use of cans that causes small airgap path.

The can loss value is measured by an indirect approach. As the first step, the total loss, including traditional iron loss, copper loss, friction loss and can loss, is measured according to the energy conversion principle. Then, friction loss as one part is separated out according to the coast down test, followed by copper loss. Due to temperature rise, copper loss at steady state is recorded. Table 3.3 comparatively shows the copper loss of both machines under rotor speeds 1 krpm and 3 krpm, with phase current levels measured by percent of the rated hysteresis 60A. In either speed, loss of the canned machine is higher and the gap by comparison is increased with a higher percent. This is due to the can loss that leads to a temperature jump on coils, and thus an increase of electrical resistivity. The gap is maximized at a higher speed 3 krpm with a full percent 99.6%.

As the next step, the can loss is obtained if iron loss is further separated out. To this end, on the standard machine with the same operation, the mentioned energy

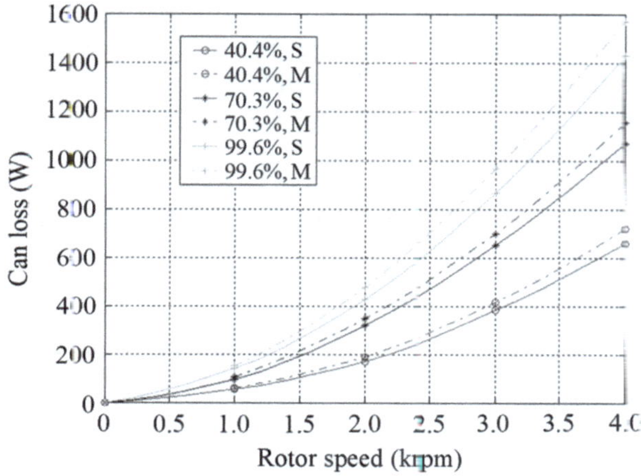

Fig. 3.12 The can loss variation with speed, under phase current levels measured by percent of the rated hysteresis limit 60A, *S* simulation, *M* measurement

conversion principle is again applied to get iron loss to be used as that of the canned machine. The obtained loss is transferrable, as from discussions in Table 3.2 that the use of cans has limited impact on eddy current loss, the main part of iron loss, both from electromagnetic and thermal point of view.

The can loss measurement at steady state operation condition is shown in Fig. 3.12. The high values of can loss indicate that DC excitation brings flux harmonics to which the non-laminated cans are sensitive. It is shown from full current level 99.6% that (1) The simulated average loss at low speed 2 krpm in Fig. 3.7 and medium speed 3 krpm in Fig. 3.9 agree with measurement; (2) Due to hardware limit, 4.5 krpm in Fig. 3.9 is not tested. However, it is predicted from the curve extension that the simulation will agree. For partial current levels 40.4% and 70.3%, the simulation and measurement agree as well.

The loss value is approximately square proportional to the rate of change of airgap flux. In Fig. 3.12, the loss inclines up drastically with speed. Meanwhile, under a given speed where excitation rises up, the loss increases as well. However, the acceleration of increase is reduced, due to magnetic saturation that hinders the change of flux. Note that there is a discrepancy by comparing simulation and measurement. In simulation model, electrical resistivity of the cans is set constant, which is referred from the average temperature measurement value at the steady state. However practically, there exist thermal sensitive regions at the axial end part of the cans, being small but causing higher temperature rise and hence locally higher electrical resistivity. The thermal end effect is not included in simulation. Therefore, simulated loss is lower and the gap is enlarged by a higher speed or current level.

In summary, the canned switched reluctance machine makes a promising solution for hydraulic pump drives. Due to characteristic use of cans, operation principles and performances make a difference. In particular the ohmic loss from cans is the principal consideration. In this section, electromagnetic features and loss due to the use of cans are analyzed, based on a salient SRM. The induced eddy current on cans is typically featured by circulation around the fluxing/defluxing pole pair with concentration at the end part. Can loss variation at different control modes shows the featured couple of peaks, and the average values rise up with rotor speed geometrically and with current excitation in a saturated way. As a direct effect, inserting a can in airgap will weaken airgap flux and harmonics, reducing eddy current loss in rotor/stator part.

Reference

1. C. Laudensack, Y. Polonskiy, D. Gerling, Measurement performance analyses of dry-running and canned switched reluctance machines, IEEE International Conference on Electrical Machines & Drives, IEMDC 2015, pp. 376–382

Chapter 4
An Analytical Model of Concentric Layer Structure for Canned Machines, Part I: Armature Coils

4.1 Model Introduction

Concentrated coils have been gaining interest for a higher energy conversion efficiency and low cost [1–3]. As to SRMs, all turns are featured by being fixed onto a single tooth. To have a comprehensive analysis, including the use of cans, an analytical model is necessary, in which the magneto-motive force (MMF) distribution is one primary consideration [4, 5]. In this chapter an analytical MMF model is proposed.

For a coil with N turns and each carrying current i, the MMF Θ, a multiplication of winding function and phase current excitation, can be written by Ampere's circuital law as

$$\Theta = Ni = \oint \vec{H}\, d\,\vec{l} \tag{4.1}$$

where Θ is closely related to magnetic field intensity H. Once the winding topology is given, (4.1) will be further improved. $N(\theta)$ is the winding function, which refers to the number of penetration distributed along the airgap circumference by a single wire into and out of the plane.

With starting point selected, the winding function $N(\theta)$ has periodicity and zero integration value through circumferential range 2π. Figure 4.1 shows an arbitrary layout of a single wire and the corresponding function respectively. The MMF is then given as

$$\Theta(\theta) = N(\theta)i \tag{4.2}$$

The MMF distribution is the prerequisite to analyze electrical machines. In this chapter, the MMF of SRMs with concentrated coils is developed, and compared with distributed coils using Fourier methods, where both the phase current excitation and

Q. Yu et al., *Analysis and Mathematical Models of Canned Electrical Machine Drives*, https://doi.org/10.1007/978-981-13-2745-2_4

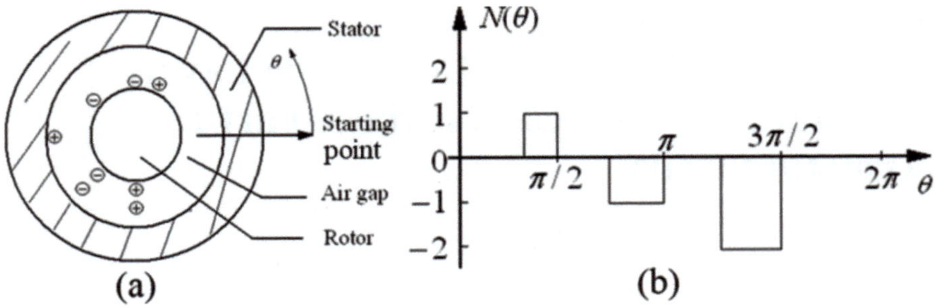

Fig. 4.1 Arbitrary layout of a single wire (**a**) and corresponding winding function (**b**)

winding topology are studied respectively. For the concentrated coils, layout of turns inside a coil is considered to analyze winding topology. Chapter organization is as follows. In Sect. 4.2, the winding function of both concentrated and distributed coil topologies are comparatively investigated. Staring from a single turn of wire, the fundamental Fourier method is introduced, with special attention on winding factor harmonics analysis. Specifically for concentrated coils, mathematical model of a single turn is first studied, followed by all turns that make a coil. Alternatively, a simplified model of a tooth concentrated coil is proposed, which considers all turns within a coil as integrity. Then, spatial distribution of 3-phase armature coils is deduced. In Sect. 4.3, phase current is modeled, with attention on DC excitation harmonics. In Sect. 4.4, MMF distribution is deduced, with different models of winding functions. In Sect. 4.5, the deduced MMF model is verified and discussed. As amendment in Sect. 4.6, the MMF distribution along axial direction is studied. As an application example in Sect. 4.7, the deduced MMF model of an SRM is used to analyze airgap flux density distribution, combining the airgap reluctance calculated from MEC models in Chap. 2. As an amendment, MMF due to permanent magnets (PMs) is deduced in Sect. 4.8.

4.2 Modeling of Winding Function

4.2.1 Fourier Approach for a Single Turn of Wire

As the first step, the winding function $N(x)$ by a single turn is considered. Figure 4.2 shows the winding function distribution along circumference x. To simplify calculation, the turn-in "+" and turn-out "–" of the wire are assumed to be concentrated and hence the waveform rises up or drops down in an escalated approach. The wire is assumed in occupation half of the period and the vertical axis is located in the middle of the wire pair. τ_p is pole pitch, and A is the amplitude that holds true within the circumference $x = \pm\tau_p/2$. Figure 4.2 in concept illustrates the winding function and if the machine has more than one pair of poles p, the waveform $N(x)$ is repeated.

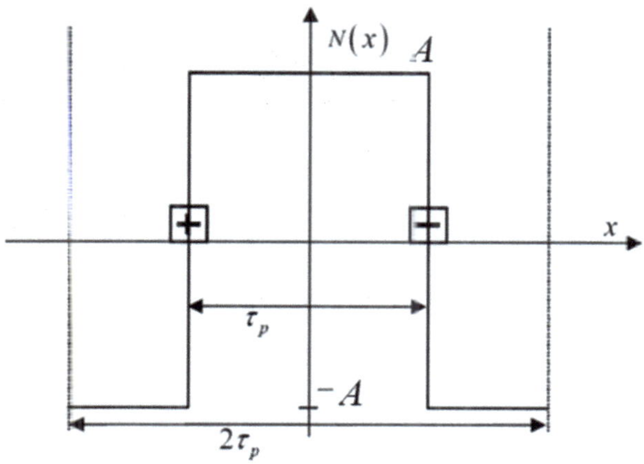

Fig. 4.2 MMF distribution of a single wire under ideal condition

As the winding function is considered a complete circulation, the distribution $N(x)$ is periodical in $2p\tau_p$. Now, a coordinate transformation is performed as follows,

$$\beta = x(\pi/\tau_p) \tag{4.3}$$

The angle β is called electrical angle, the function $N(x)$ is periodical in $0 \leq x < 2\tau_p$, and therefore $N(\beta)$ is periodical in $0 \leq \beta < 2\pi$, with

$$\alpha = \frac{\beta}{p} = x\frac{\pi}{p\tau_p} \tag{4.4}$$

where α is called mechanical angle. If the electrical conditions repeat pole pairs p-times along the circumference, the angular range is $0 \leq \beta < 2\pi p$ when rotor with angle α passes one period $[0, 2\pi]$. Now the winding function $N(x)$ in (4.5) is analyzed to get Fourier series. The general equation is written as (for $p = 1$)

$$N(x) = \frac{a_0}{2} + \sum_{v} a_v \cos(vx) + \sum_{v} b_v \sin(vx) \tag{4.5}$$

where the parameter v is called circumferential harmonic. Due to the symmetry along vertical axis, the average of $N(x)$ is zero and hence $a_0 = 0$ holds true. Coefficients in (4.5) are further written as

$$a_v = -\frac{1}{v\pi}\sum_{\lambda} c \sin(v\lambda), \quad v = 1, 3, 5, \ldots \tag{4.6}$$

$$b_\nu = +\frac{1}{\nu\pi}\sum_{\lambda} c\cos(\nu\lambda), \quad \nu = 1, 3, 5, \ldots \tag{4.7}$$

where λ is discretization harmonic and c is a constant coefficient. According to Fig. 4.2, within the period of $\lambda = \pm\pi/2$ and $c = \pm N$, the following is obtained,

$$a_\nu = -\frac{1}{\nu\pi}\left(\sin\left(-\nu\frac{\pi}{2}\right) - \sin\left(-\nu\frac{\pi}{2}\right)\right) = \frac{2}{\pi\nu}\sin\left(\frac{\pi}{2}\nu\right), \quad \nu = 1, 2, 3, \ldots \tag{4.8}$$

$$b_\nu = +\frac{1}{\nu\pi}\left(\cos\left(-\nu\frac{\pi}{2}\right) - \cos\left(-\nu\frac{\pi}{2}\right)\right) = 0 \tag{4.9}$$

In (4.8), the term $\sin(\nu(\pi/2))$ is introduced, which describes the sum in an infinite form. In (4.9), $b_\nu = 0$ always holds true, as the function in Fig. 4.2 is symmetrical to vertical axis.

When number of pole pairs p is considered, it must be taken into account the multiple complete circulations around bore circumference. The frequencies of Fourier coefficients will be increased by the factor p, whereas the amplitudes of Fourier coefficients remain unchanged. The jump height of function $N(x)$ should also stay the same. Therefore, (4.8) and (4.9) are further written as

$$a_\nu = \frac{2}{\pi\nu}\sin\left(\frac{\pi}{2}\frac{\nu}{p}\right), \quad \nu/p = 1, 2, 3, \ldots \tag{4.10}$$

$$b_\nu = 0, \quad \nu/p = 1, 2, 3, \ldots \tag{4.11}$$

$$\sin\left(\frac{\pi}{2}\frac{\nu}{p}\right) = \begin{cases} 1 & \text{for} \quad \nu/p = 1, 5, 9 \\ -1 & \text{for} \quad \nu/p = 3, 7, 11 \\ 0 & \text{for} \quad \nu/p = 2, 4, 6 \end{cases} \tag{4.12}$$

With (4.5), (4.6), (4.7), (4.8), (4.9), (4.10), (4.11) and (4.12), the following equation is obtained as

$$N(\beta) = \frac{2}{\pi}\sum_{\nu}\frac{\sin\left(\frac{\nu}{p}\frac{\pi}{2}\right)}{\nu}\cos\left(\frac{\nu}{p}\beta\right), \quad \nu/p = 1, 3, 5, \ldots \tag{4.13}$$

Alternatively, $N(x)$ by ν-th harmonic is written as

$$^{\nu}N(\beta) = \frac{2}{\pi}\frac{\sin\left(\frac{\nu}{p}\frac{\pi}{2}\right)}{\nu}\cos\left(\frac{\nu}{p}\beta\right), \nu/p = 1, 3, 5, \ldots \tag{4.14}$$

Further (4.14) in complex form is written as

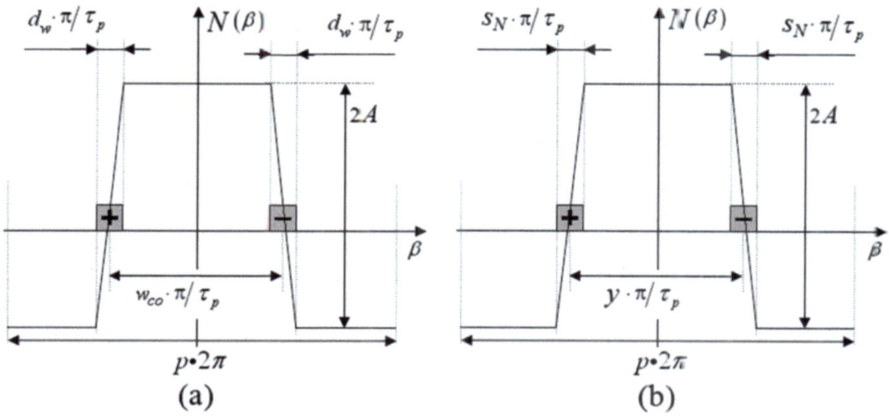

Fig. 4.3 Improved models of a single turn, (**a**) concentrated, (**b**) distributed

$$^{\nu}N(\beta) = \mathrm{Re}\left(\frac{2}{\pi}\frac{\sin\left(\frac{\nu}{p}\frac{\pi}{2}\right)}{\nu}e^{-j\frac{\nu}{p}\beta}\right) = \frac{2}{\pi\nu}\sin\left(\frac{\nu}{p}\frac{\pi}{2}\right)\mathrm{Re}\left(e^{-j\frac{\nu}{p}\beta}\right), \quad \nu/p = 1, 3, 5, \ldots \tag{4.15}$$

Considering negative harmonics, the coefficient 1/2 is introduced into (4.13), (4.14) and (4.15), the complex representation for the ν-th harmonic along the complete circumference x is then rewritten as

$$^{\nu}N(\beta) = \frac{\sin\left(\frac{\nu}{p}\frac{\pi}{2}\right)}{\pi\nu}e^{-j\frac{\nu}{p}\beta}, \quad \nu/p = \pm 1, \pm 3, \pm 5, \ldots \tag{4.16}$$

Based on above deduction, Fourier discretization on both concentrated and distributed coils is studied. Improvements are shown in Fig. 4.3, in which compared with Fig. 4.2, the rate of rising or dropping during turn-in and turn-out ranges is taken to model actual coil excitation. Hence, the waveform in Fig. 4.3 is in a trapezoidal form instead of a rectangle. In addition, the electrical angle β is applied as circumferential variation to replace x. By comparison, both waveforms are quite similar and the difference lies in the arrangement of coils. In Fig. 4.3a for concentrated coils, diameter of a turn d_w and circumferential span of a coil w_{co} are designated, while in Fig. 4.3b for distributed coils, pole pitch y and s_N slot opening width characterize the coil geometrical features. Both figures are of a period $2p\pi$ and symmetrical to vertical axis. The term (π/τ_p) converts corresponding lengths into electrical degree β.

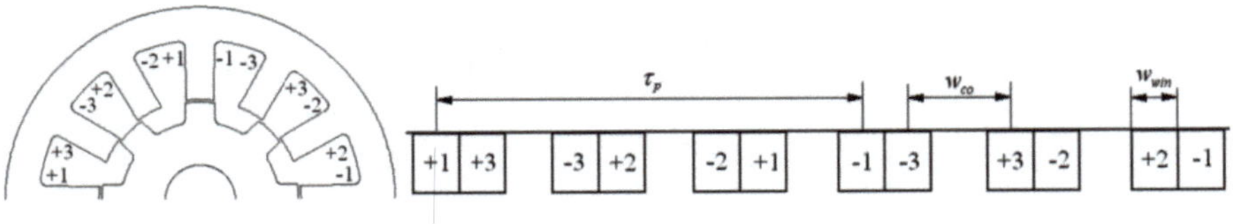

Fig. 4.4 Definition of concentrated armature coils

Fig. 4.5 Geometrical specifications of a single turn for tooth concentrated coils

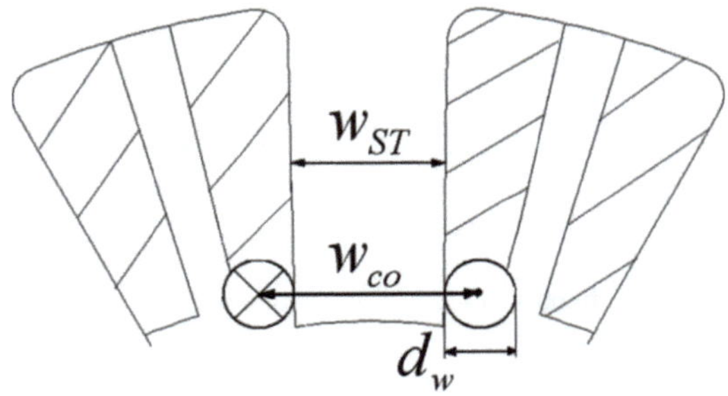

4.2.2 *Model of a Tooth Concentrated Coil with Wire Layout (Model 1)*

This model is characterized by investigating the arrangement of wires within a coil. Sketch of the studied 12/8, 3-phase SRM in this chapter, with geometric specifications in Table 2.8, is shown in Fig. 4.4. The ordinal numbers 1,2,3 stand for coils from different phases and the prefix signs for current direction. Concentrated coils are wound around a tooth through adjacent slots. There are altogether 4 coils in one phase with an interval of 90° in mechanical degree. Note that the mechanical angles with regard to winding topologies should be always headed with p to be converted into electrical degrees.

Regardless of curvature, coils are unfolded and shown, in which some measurements are defined. τ_p refers to distance between adjacent stator poles of a phase. $\tau_p = 2\pi R_{ST}/4 = \pi R_{ST}/2$, where R_{ST} refers to radius from center to top of a stator tooth. w_{win} means circumferential thickness of one side of a coil, and due to the topology of tooth concentrated coils, w_{win} has difference compared with slot pitch s_N for the distributed coils. Due to the gap between a couple of coil sides in one slot, w_{win} value is less than half of s_N. Alternatively, w_{co} is the width between two sides of a coil, which in analogy has difference to τ_N as well. w_{co} relates with w_{win}, and that is, let $w_{co} = w_{win} + w_{ST}$, where w_{ST} is the width of stator teeth. Usually teeth are slightly wedged and the average width value is used. Figure 4.5 illustrates the geometrical relations. A wire can be reduced collectively to 2 dots on each side of a tooth with indication of current direction. The dots are represented by two circular blocks and the geometrical relation between d_w and w_{co} are shown.

Based on Fig. 4.3a, Fourier discretization is performed. Note that the winding function $N(\beta)$ now consists of several straights with slope values. According to definition of a pole pair that describes the rotating speed ratio between electrical excitation and mechanical rotor, for a 3-phase 12/8 SRM with number of rotor poles $N_r = 8$, one phase current will finish one period when one rotor pole moves $2\pi/N_r$. Therefore for SRMs the equivalent number of pole pairs $p = N_r$. Hence, the angular range of electrical angle β is $[-p\pi, p\pi] = [-8\pi, 8\pi]$. Also note that in such a case only one wire of a coil is modeled. In the following, each straight line is deduced, which is distinguished by having different slope values s'.

$$-4\pi < \beta < -\frac{w_{co}}{\tau_p}\frac{\pi}{2} - \frac{d_w}{\tau_p}\frac{\pi}{2}: \quad s' = 0 \tag{4.17}$$

$$-\frac{w_{co}}{\tau_p}\frac{\pi}{2} - \frac{d_w}{\tau_p}\frac{\pi}{2} < \beta < -\frac{w_{co}}{\tau_p}\frac{\pi}{2} + \frac{d_w}{\tau_p}\frac{\pi}{2}: \quad s' = \frac{\tau_p}{d_w\pi} \tag{4.18}$$

$$-\frac{w_{co}}{\tau_p}\frac{\pi}{2} + \frac{d_w}{\tau_p}\frac{\pi}{2} < \beta < \frac{w_{co}}{\tau_p}\frac{\pi}{2} - \frac{d_w}{\tau_p}\frac{\pi}{2}: \quad s' = 0 \tag{4.19}$$

$$\frac{w_{co}}{\tau_p}\frac{\pi}{2} - \frac{d_w}{\tau_p}\frac{\pi}{2} < \beta < \frac{w_{co}}{\tau_p}\frac{\pi}{2} + \frac{d_w}{\tau_p}\frac{\pi}{2}: \quad s' = -\frac{\tau_p}{d_w\pi} \tag{4.20}$$

$$\frac{w_{co}}{\tau_p}\frac{\pi}{2} + \frac{d_w}{\tau_p}\frac{\pi}{2} < \beta < 4\pi: \quad s' = 0 \tag{4.21}$$

The waveform in Fig. 4.3a should be transformed into Fourier series. According to Appendix A.1, the coefficients is simplified as

$$a_v = -\frac{1}{v^2\pi}\sum_{\lambda'} s' \cos(\lambda' v/p), \quad v = 1, 2, 3, \ldots \tag{4.22}$$

$$b_v = -\frac{1}{v^2\pi}\sum_{\lambda'} s' \sin(\lambda' v/p), \quad v = 1, 2, 3, \ldots \tag{4.23}$$

where λ' is the point of discontinuity in $N'(\beta)$. The coefficient can be written as

$$a_v = \frac{2i}{v\pi}\sin\left(\frac{v}{p}\frac{w_{co}}{\tau_p}\frac{\pi}{2}\right)\mathrm{si}\left(\frac{v}{p}\frac{d_w}{\tau_p}\frac{\pi}{2}\right) \tag{4.24}$$

$$b_v = 0 \tag{4.25}$$

where the function "si" is defined as

$$\mathrm{si}(x) = \sin(x)/x \tag{4.26}$$

The winding function $N(\beta)$ of a single turn can be expressed as

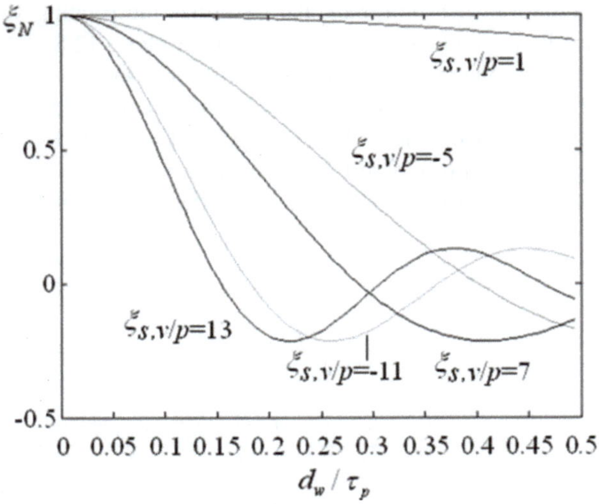

Fig. 4.6 The wire thickness factor variation by circumferential harmonic ν

$$N(\beta) = \sum_{\nu} \frac{2}{\nu\pi} \sin\left(\frac{\nu}{p}\frac{w_{co}}{\tau_p}\frac{\pi}{2}\right) \mathrm{si}\left(\frac{\nu}{p}\frac{d_w}{\tau_p}\frac{\pi}{2}\right) \cos\left(\frac{\nu}{p}\beta\right), \quad \nu/p = 1, 2, 3\ldots \tag{4.27}$$

$$^{\nu}\xi_N = \mathrm{si}\left(\frac{\nu}{p}\frac{d_w}{\tau_p}\frac{\pi}{2}\right) \tag{4.28}$$

$$^{\nu}\xi_S = \sin\left(\frac{\nu}{p}\frac{w_{co}}{\tau_p}\frac{\pi}{2}\right) \tag{4.29}$$

The factor $^{\nu}\xi_N$ here describes circumferential width of a wire, where the superscript ν indicates circumferential harmonics. In some literatures the term "slot opening factor" is used in distributed or fractional concentrated windings in which the width of a coil is alternatively modeled the same as that of the slot for simplicity. Here this factor helps to determine thickness of coils featured by wire diameter d_w. For concentrated windings the name "wire thickness factor" is used and the waveform is shown in Fig. 4.6, indicating a wire of a higher diameter helps to eliminate harmonics. The short pitch factor $^{\nu}\xi_S$ is defined as the ratio w_{co}/τ_p and the waveform is shown in Fig. 4.7. The ratio is expected to be higher to make the fundamental order more prominent, and meantime w_{co} should be selected properly to suppress harmonics.

Alternatively, (4.27) can be now written as

$$N(\beta) = \sum_{\nu} \frac{2}{\nu\pi} {}^{\nu}\xi_N {}^{\nu}\xi_S \cos\left(\frac{\nu}{p}\beta\right), \quad \nu/p = 1, 2, 3\ldots \tag{4.30}$$

(4.30) along the complete circumference range β is written as

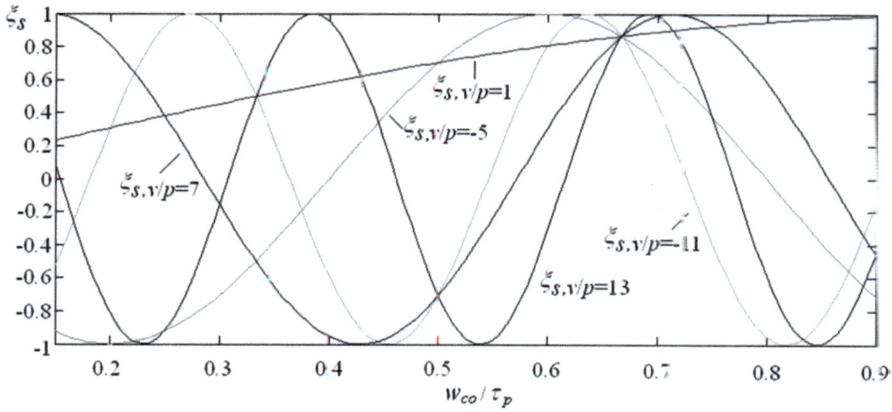

Fig. 4.7 The short pitch factor variation by circumferential harmonic v

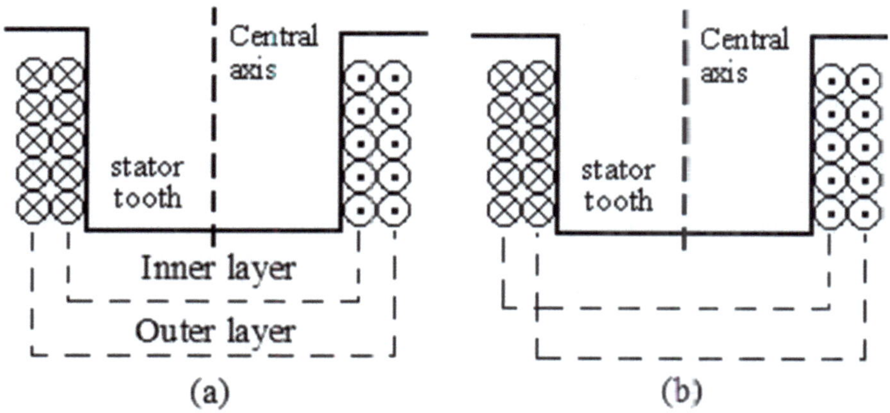

Fig. 4.8 Wire distribution overview and definition of a layer, (**a**) structural overview, (**b**) regrouping

$$N(\beta) = \sum_{v} \frac{1}{v\pi} {}^{v}\xi_N {}^{v}\xi_S \cos\left(\frac{v}{p}\beta\right), \quad v/p = \pm 1, \pm 2, \pm 3 \ldots \tag{4.31}$$

In the following, modeling of wire arrangement within a col. is studied, based on a single turn. As an application example, an armature coil consisting of N_p turns is studied and a two-layer topology is shown. In Fig. 4.8, each turn in a coil is presented with "×" and "•", indicating current direction. Figure 4.8 shows the actual way of windage and its regrouping. In Fig. 4.8a, all turns revolve around the central axis to which both the inner and outer layers are symmetrical. Layers are regrouped to have the same pitch in Fig. 4.8b. The equivalence holds true as the same MMF can be generated. Note that symmetry axes of layers after regrouping do not necessarily fall

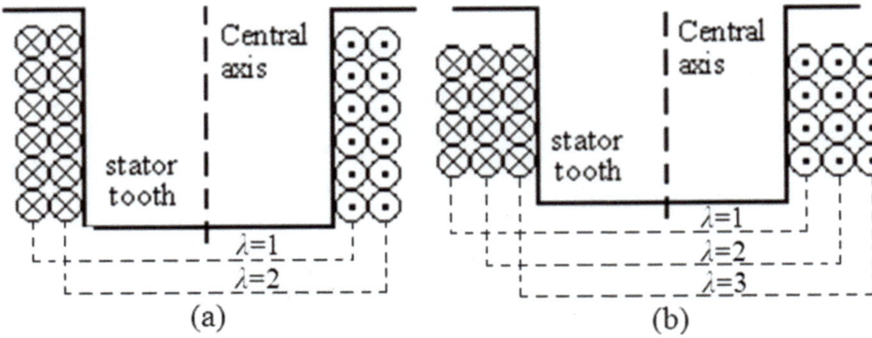

Fig. 4.9 Definition of a layer of wires within a coil, (**a**) $S = 2$, (**b**) $S = 3$

onto the central axis. As a result when describing the wire distribution, there will be a circumferential displacement β_S for a layer in the winding function equation.

As a typical structure to be modeled, assuming all turns are evenly wound along the central axis, there are altogether 12 turns in the array, e.g. could be 2×6, or 3×4 and we assume $N_p = 12$. The MMF excited via each turn is seen identical and all turns in one column of the array are grouped as a layer. Hence, the layout of a coil is described in terms of layers. In Fig. 4.9a, there are 2 layers $\lambda = 1{,}2$ ($S = 2$) while in Fig. 4.9b, there are 3 layers $\lambda = 1{,}2{,}3$ ($S = 3$) and each layer has the same distance between the "×" and "•" side.

The winding function from Layer 1 is first described and then distribution of all the other layers is added. When modeling one layer, the wires are numbered with χ. For example for $\lambda = 1$ in Fig. 4.9a, $\chi = 1{,}2{,}\ldots{,}6$ from top to bottom of the tooth and in analogy for $\lambda = 2$, $\chi = 7{,}8{,}\ldots{,}12$ are defined. Modeling of a layer starts from a turn, in which a similar method from Fig. 4.3a is used that all the geometrical relations can be deducted if wire diameter d_w and width of teeth w_{ST} are known. Note that the pole pitch τ_p should be changed to $\tau_{p,\chi}$, which in analogy to pole pitch refers to quarter of the circumference measured by radius from center point to the χ-th turn and is written as

$$\tau_{p,\chi} = 2\pi(R_{ST} + \chi d_w)/4 = \tau_p + \pi\chi d_w/2 \tag{4.32}$$

where R_{ST} refers to radius of stator teeth. Similar to (4.28) and (4.29), winding factors are rewritten as (Note that subscript χ is added to indicate the ordinal number of turns)

$$^{v}\xi_{S,\chi} = \sin\left(\frac{v}{p}\frac{w_{ST} + 2d_w}{\tau_{p,\chi}}\frac{\pi}{2}\right), \quad v/p = 1, 2, 3\ldots \tag{4.33}$$

$$^{v}\xi_{N,\chi} = \mathrm{si}\left(\frac{v}{p}\frac{d_w}{\tau_{p,\chi}}\frac{\pi}{2}\right), \quad v/p = 1, 2, 3\ldots \tag{4.34}$$

Note that the subscript S in $^{v}\xi_{N,\chi}$ for short pitch and the variable S in the expression for number of layers in a coil are not related. By sum of all turns, in analogy to (4.30), the winding function of a layer is written as

$$^{v}N(\beta) = \sum_{\chi=1}^{S} \frac{2}{v\pi} {}^{v}\xi_{N,\chi} {}^{v}\xi_{S,\chi} \cos\left(\frac{v}{p}(\beta - (d_w/2)(\pi/\tau_{p,\chi}))\right), \quad v/p = 1, 2, 3\ldots \tag{4.35}$$

For S-th layer, each wire has a translational move of $((S-1)/2)d_w$, with the term $(\pi/\tau_{p,\chi})$ being converted into electrical angles. There are (N_p/s) turns and therefore, according to the mentioned circumferential displacement due to sum of turns, the angular shift of S-th layer β_S is deduced as

$$\begin{aligned} \beta_S &= \left(\frac{N_p}{S}\right)^{-1} \sum_{\chi=1}^{N_p/S} ((S-1)/2) d_w (\pi/\tau_{p,\chi}) \\ &= \frac{S}{2N_p}(S-1) d_w \pi \sum_{\chi=1}^{N_p/S} (\tau_{p,\chi})^{-1} \\ &= \frac{2S^2\pi(S-1)d_w}{N_p(4S\tau_p + \pi N_p w_{SS})} \end{aligned} \tag{4.36}$$

where S is the total number of layers, N_p is number of turns per pole, w_{SS} is the width of stator slot neglecting curvature. Note that the coefficient $(N_p/s)^{-1}$ ahead of the sum means to average the total shift of all wires. Then, the winding function by a layer of wires is written as

$$^{v}N(\beta) = \sum_{\chi=1}^{N_p/S} \frac{2}{v\pi} {}^{v}\xi_{N,\chi} {}^{v}\xi_{S,\chi} \cos\left(\frac{v}{p}(\beta - \beta_S)\right), \quad v/p = 1, 2, 3\ldots \tag{4.37}$$

(4.37) can be further written as

$$^{v}N(\beta) = \frac{2}{v\pi} {}^{v}\xi_{NS} \cos\left(\frac{v}{p}(\beta - \beta_S)\right), \quad v/p = 1, 2, 3\ldots \tag{4.38}$$

$$ {}^{v}\xi_{NS} = \sum_{\chi=1}^{N_p/S} {}^{v}\xi_{N,\chi} {}^{v}\xi_{S,\chi} \tag{4.39} $$

${}^{v}\xi_{NS}$ combines thickness factor ${}^{v}\xi_{N}$ and short pitch factor ${}^{v}\xi_{S}$ of each wire. Then, sum of all layers is performed to form a coil. To make the sum, there is a circumferential shift of the γ-th layer based on a baseline layer. According to Fig. 4.9, the shift is written as

$$ \beta_{\gamma} = -(\gamma - 1)d_w\left(\pi/\tau_{p,\gamma}\right) \tag{4.40} $$

where the γ-th layer is required to move a circumferential length of $(\gamma-1)d_w$, which is converted into electrical angle via the term $(\pi/\tau_{p,\gamma})$. The minus sign indicate clockwise direction. In analogy to $\tau_{p,\chi}$, $\tau_{p,\gamma}$ refers to quarter of the circumference measured by radius from center point to midpoint of γ-th layer, where the subscript γ in $\tau_{p,\gamma}$ indicates the linkage with number layers. $\tau_{p,\gamma}$ is defined as

$$ \begin{aligned} \tau_{p,\gamma} &= 2\pi\left(R_{ST} + \frac{N_p}{2S}d_w\right)/4 = \pi R_{ST}/2 + \pi\frac{N_p}{2S}d_w/2 \\ &= \tau_p + \pi\frac{N_p}{4S}d_w \end{aligned} \tag{4.41} $$

The sum of layers is performed by sum of unit vectors based on the following equation, where j is the unit vector,

$$ \sum_{\mu=1}^{n} e^{j\mu\lambda} = \frac{\sin\left(n\frac{\lambda}{2}\right)}{\sin\left(\frac{\lambda}{2}\right)} e^{j(n+1)\frac{\lambda}{2}} \tag{4.42} $$

In (4.42), by replacing λ with β_{γ} from (4.40), the sum of all layers leads to

$$\begin{aligned}
\sum_{\gamma=1}^{S} e^{j\frac{v}{p}\beta_\gamma} &= \sum_{\gamma=1}^{S} e^{j\frac{v}{p}\left(-\pi d_w(\gamma-1)/\tau_{p,\gamma}\right)} \\
&= e^{j\frac{v}{p}\left(\pi d_w/\tau_{p,\gamma}\right)} \sum_{\gamma=1}^{S} e^{j\frac{v}{p}\left(-\pi d_w \gamma/\tau_{p,\gamma}\right)} \\
&= e^{j\frac{v}{p}\left(\pi d_w/\tau_{p,\gamma}\right)} \frac{\sin\left(\frac{S}{2}\frac{v}{p}\frac{\pi d_w}{\tau_{p,\gamma}}\right)}{\sin\left(\frac{1}{2}\frac{v}{p}\frac{\pi d_w}{\tau_{p,\gamma}}\right)} e^{j(S+1)\frac{1}{2}\frac{v}{p}\frac{-\pi d_w}{\tau_{p,\gamma}}} \\
&= \frac{\sin\left(S\frac{v}{p}\frac{\pi d_w}{2\tau_{p,\gamma}}\right)}{\sin\left(\frac{v}{p}\frac{\pi d_w}{2\tau_{p,\gamma}}\right)} e^{j\frac{v}{p}\left(-(S-1)\frac{\pi d_w}{2\tau_{p,\gamma}}\right)} \\
&= S\frac{\sin\left(S\frac{v}{p}\frac{\pi d_w}{2\tau_{p,\gamma}}\right)}{S\sin\left(\frac{v}{p}\frac{\pi d_w}{2\tau_{p,\gamma}}\right)} e^{j\frac{v}{p}\left(-(S-1)\frac{\pi d_w}{2\tau_{p,\gamma}}\right)} \\
&= S^{v}\xi_Z e^{j\frac{v}{p}\left(-(S-1)\frac{\pi d_w}{2\tau_{p,\gamma}}\right)}, \quad v/p = \pm 1, \pm 2, \pm 3\ldots
\end{aligned} \tag{4.43}$$

$$^{v}\xi_Z = \frac{\sin\left(S\frac{v}{p}\frac{\pi d_w}{2\tau_{p,\gamma}}\right)}{S\sin\left(\frac{v}{p}\frac{\pi d_w}{2\tau_{p,\gamma}}\right)}, \quad v/p = \pm 1, \pm 2, \pm 3\ldots \tag{4.44}$$

The distribution factor $^{v}\xi_Z$ describes winding function determined by distribution of wires within a coil, a characteristic feature which offers theoretical guidance on harmonic analysis for tooth concentrated windings. In some literatures, the short pitch factor and distribution factor are collectively defined as a means of relating harmonic components of a winding of arbitrary distribution to a common reference, which helps to eliminate harmonics. Therefore according to (4.31), (4.36), (4.39) and (4.44), the winding function of a coil by sum of all layers leads to (where complex form is underlined)

$$\underline{N}(\beta) = \sum_{v} \frac{S}{v\pi} {}^{v}\xi_{NS} {}^{v}\xi_Z \cos\left(\frac{v}{p}(\beta - \beta_S)\right) e^{j\frac{v}{p}\left(-(S-1)\frac{\pi d_w}{2\tau_{p,\gamma}}\right)}, \quad v/p = \pm 1, \pm 2, \pm 3\ldots \tag{4.45}$$

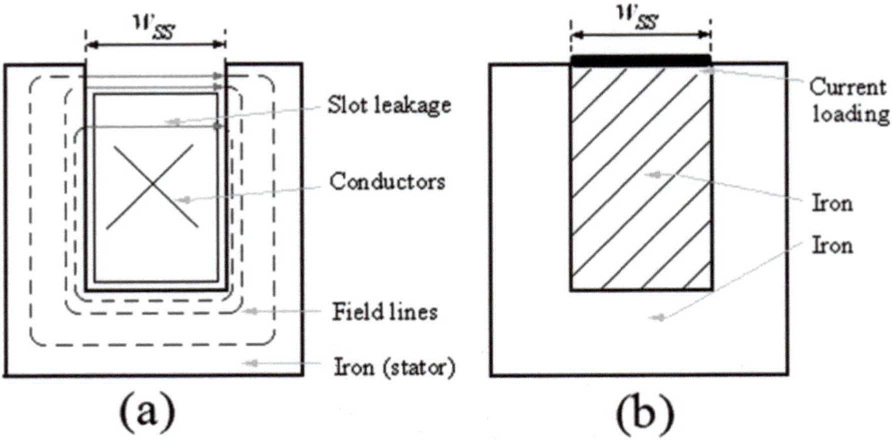

Fig. 4.10 Transfer a coil onto slot surface, (**a**) a slot and coil, (**b**) equivalent transfer

4.2.2.1 Simplified Model of a Tooth Concentrated Coil (Model 2)

Instead of investigating wire layout, this model is simplified by taking all wires within a coil as integrity. That is, the internal structure is neglected and it is assumed that all wires of a coil are uniformly distributed in the slot and only the number of turns N_p is necessary. The coil in slot can be transferred onto top of the slot. Figure 4.10a shows one slot filled with a coil and the excited flux field circulates via stator iron and airgap width w_{SS}. The relative permeability of iron is assumed being very large against air. Then in Fig. 4.10b, a current loading responsible for this field can be switched evenly onto surface of the slot, meanwhile with slot inside filled with homogenous iron.

As to an actual tooth concentrated coil, the slot is separated into halves and both parts are filled with a coil side that has reverse current excitation (Fig. 4.11a). The simplification is accordingly shown in Fig. 4.11b, where both coil sides are transferred to the slot surface and each is with a width $w_{SS}/2$.

The winding function is investigated concerning such transfer. In Fig. 4.12, a couple of rectangular blocks with prefix signs correspond to coil sides in Fig. 4.11b. Therefore, diameter of a wire d_w in is replaced by half of the slot opening width $w_{SS}/2$ compared with Fig. 4.3a, and width of a coil w_{co} in this occasion equals to $(w_{ST} + w_{SS}/2)$ according to the geometrical relation in Fig. 4.5. Similarly, (π/τ_p) converts all measurements into electrical degrees.

In analogy, the mathematical expression of the winding function can be written as

$$N(\beta) = \sum_{v} \frac{2}{v\pi} {}^{v}\xi_N {}^{v}\xi_S \cos\left(\frac{v}{p}\beta\right), \quad v/p = 1, 2, 3 \ldots \tag{4.46}$$

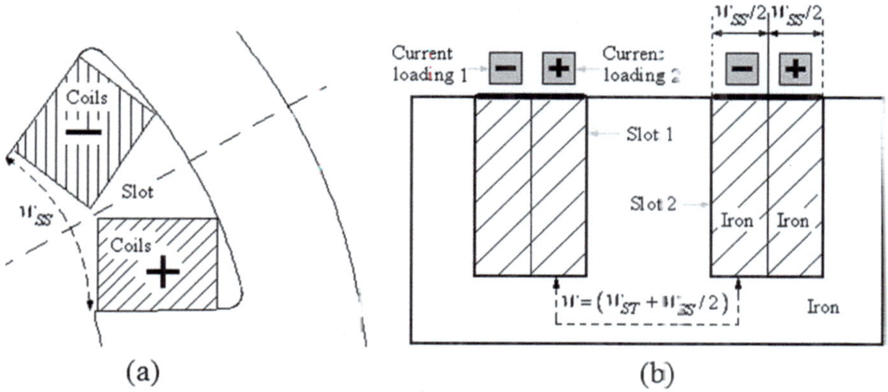

Fig. 4.11 Modeling of tooth concentrated coils, (**a**) coil distribution, (**b**) equivalent transfer

Fig. 4.12 The winding function curve of a tooth concentrated coil using a simplified approach

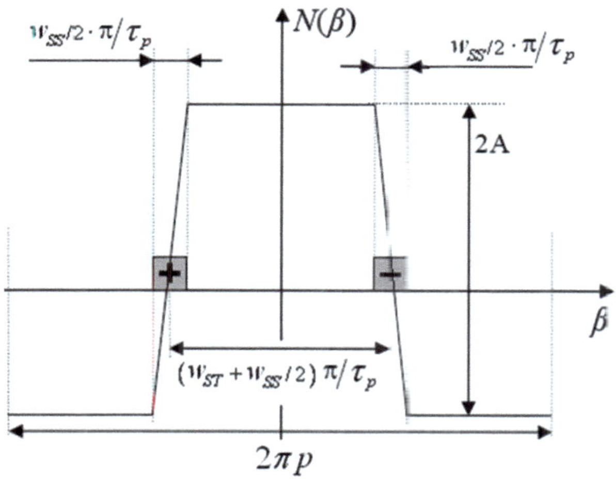

$$^{v}\xi_N = \mathrm{si}\left(\frac{v}{p}\frac{w_{SS}}{2\tau_p}\frac{\pi}{2}\right) \tag{4.47}$$

$$^{v}\xi_S = \sin\left(\frac{v}{p}\frac{w}{\tau_p}\frac{\pi}{2}\right) \tag{4.48}$$

where $w = (w_{ST} + w_{SS}/2)$ is used.

4.2.3 Model of One Coil in Distributed Topology

For the distributed winding topology, there is only one coil side in each slot. In analogy to Sect. 4.2.3, all wires within a coil are taken as integrity. In analogy, s_N is slot opening width, τ_p is pole pitch and the coil width w_{co} is replaced by y. According to Fig. 4.5a, we assume that there are also N_p turns that are the same and concentrated in one slot. Instead of N_p, we use N_c for distributed coils, which means the number of turns within a coil. First, let $p = 1$, we get

$$-\pi < \beta < -\frac{y}{\tau_p}\frac{\pi}{2} - \frac{s_N}{\tau_p}\frac{\pi}{2}: \quad s' = 0 \tag{4.49}$$

$$-\frac{y}{\tau_p}\frac{\pi}{2} - \frac{s_N}{\tau_p}\frac{\pi}{2} < \beta < -\frac{y}{\tau_p}\frac{\pi}{2} + \frac{s_N}{\tau_p}\frac{\pi}{2}: \quad s' = N_c\frac{\tau_p}{s_N\pi} \tag{4.50}$$

$$-\frac{y}{\tau_p}\frac{\pi}{2} + \frac{s_N}{\tau_p}\frac{\pi}{2} < \beta < \frac{y}{\tau_p}\frac{\pi}{2} - \frac{s_N}{\tau_p}\frac{\pi}{2}: \quad s' = 0 \tag{4.51}$$

$$\frac{y}{\tau_p}\frac{\pi}{2} - \frac{s_N}{\tau_p}\frac{\pi}{2} < \beta < \frac{y}{\tau_p}\frac{\pi}{2} + \frac{s_N}{\tau_p}\frac{\pi}{2}: \quad s' = -N_c\frac{\tau_p}{s_N\pi} \tag{4.52}$$

$$\frac{y}{\tau_p}\frac{\pi}{2} + \frac{s_N}{\tau_p}\frac{\pi}{2} < \beta < \pi: \quad s' = 0 \tag{4.53}$$

Likewise, we get (for $p = 1$)

$$a_v = -\frac{1}{v^2\pi}\sum_{\lambda'} s' \cos(v\lambda'), \quad v = 1, 2, 3, \ldots \tag{4.54}$$

$$b_v = -\frac{1}{v^2\pi}\sum_{\lambda'} s' \sin(v\lambda'), \quad v = 1, 2, 3, \ldots \tag{4.55}$$

where λ' is the point of discontinuity for the function $N'(\beta)$. $b_v = 0$ holds true.

For a_v we get,

$$
\begin{aligned}
a_\nu &= -\frac{1}{\nu^2\pi}\sum_{\lambda'} s'\cos(\nu\lambda'), \quad \nu = 1,2,3,\ldots \\
&= -\frac{1}{\nu^2\pi}\begin{pmatrix} N_c\frac{\tau_p}{s_N\pi}\cos\left(\nu\left(-\frac{y}{\tau_p}\frac{\pi}{2}-\frac{s_N}{\tau_p}\frac{\pi}{2}\right)\right) - N_c\frac{\tau_p}{s_N\pi}\cos\left(\nu\left(-\frac{y}{\tau_p}\frac{\pi}{2}+\frac{s_N}{\tau_p}\frac{\pi}{2}\right)\right) \\ -N_c\frac{\tau_p}{s_N\pi}\cos\left(\nu\left(\frac{y}{\tau_p}\frac{\pi}{2}-\frac{s_N}{\tau_p}\frac{\pi}{2}\right)\right) + N_c\frac{\tau_p}{s_N\pi}\cos\left(\nu\left(\frac{y}{\tau_p}\frac{\pi}{2}+\frac{s_N}{\tau_p}\frac{\pi}{2}\right)\right) \end{pmatrix} \\
&= -\frac{1}{\nu^2\pi}N_c\frac{\tau_p}{s_N\pi}\begin{pmatrix} \cos\left(\nu\left(-\frac{y}{\tau_p}\frac{\pi}{2}-\frac{s_N}{\tau_p}\frac{\pi}{2}\right)\right) - \cos\left(\nu\left(-\frac{y}{\tau_p}\frac{\pi}{2}+\frac{s_N}{\tau_p}\frac{\pi}{2}\right)\right) \\ -\cos\left(\nu\left(\frac{y}{\tau_p}\frac{\pi}{2}-\frac{s_N}{\tau_p}\frac{\pi}{2}\right)\right) + \cos\left(\nu\left(\frac{y}{\tau_p}\frac{\pi}{2}+\frac{s_N}{\tau_p}\frac{\pi}{2}\right)\right) \end{pmatrix}
\end{aligned} \tag{4.56}
$$

With $\cos(x) = \cos(-x)$, we get

$$
\begin{aligned}
a_\nu &= -\frac{1}{\nu^2\pi}\sum_{\lambda'} s'\cos(\nu\lambda'), \quad \nu = 1,2,3,\ldots \\
&= -\frac{2N_c\tau_p}{\nu^2\pi^2 s_N}\left(\cos\left(\nu\left(\frac{y}{\tau_p}\frac{\pi}{2}+\frac{s_N}{\tau_p}\frac{\pi}{2}\right)\right) - \cos\left(\nu\left(\frac{y}{\tau_p}\frac{\pi}{2}-\frac{s_N}{\tau_p}\frac{\pi}{2}\right)\right)\right)
\end{aligned} \tag{4.57}
$$

With $\cos(x \pm y) = \cos(x)\cos(y) \mp \sin(x)\sin(y)$, a_ν is further written as

$$
\begin{aligned}
a_\nu &= -\frac{2N_c\tau_p}{\nu^2\pi^2 s_N}\left(-2\sin\left(\nu\frac{y}{\tau_p}\frac{\pi}{2}\right)\sin\left(\nu\frac{s_N}{\tau_p}\frac{\pi}{2}\right)\right) \\
&= \frac{2N\tau_p}{\nu\pi}\sin\left(\nu\frac{y}{\tau_p}\frac{\pi}{2}\right)\mathrm{si}\left(\nu\frac{s_N}{\tau_p}\frac{\pi}{2}\right), \quad \nu = 1,2,3,\ldots
\end{aligned} \tag{4.58}
$$

Then, the winding function $N(\beta)$ is written as

$$
\begin{aligned}
N(\beta) &= \sum_\nu \frac{2N_c}{\nu\pi}\sin\left(\nu\frac{y}{\tau_p}\frac{\pi}{2}\right)\mathrm{si}\left(\nu\frac{s_N}{\tau_p}\frac{\pi}{2}\right)\cos(\nu\beta), \\
\nu &= 1,2,3,\ldots
\end{aligned} \tag{4.59}
$$

When taking multiple pole pairs, (4.59) is further written as

$$
\begin{aligned}
N(\beta) &= \sum_\nu \frac{2N_c}{\nu\pi}\sin\left(\frac{\nu}{p}\frac{y}{\tau_p}\frac{\pi}{2}\right)\mathrm{si}\left(\frac{\nu}{p}\frac{s_N}{\tau_p}\frac{\pi}{2}\right)\cos\left(\frac{\nu}{p}\beta\right), \\
\nu/p &= 1,2,3,\ldots
\end{aligned} \tag{4.60}
$$

With slot opening factor ${}^\nu\xi_N$ and short pitch factor ${}^\nu\xi_S$ respectively written as

$$ {}^{\nu}\xi_N = \mathrm{si}\left(\frac{\nu}{p}\frac{s_N}{\tau_p}\frac{\pi}{2}\right), \quad \nu/p = 1, 2, 3\ldots \tag{4.61} $$

$$ {}^{\nu}\xi_S = \sin\left(\frac{\nu}{p}\frac{y}{\tau_p}\frac{\pi}{2}\right), \quad \nu/p = 1, 2, 3\ldots \tag{4.62} $$

Compared to (4.28) and (4.29) the term (w_{co}/τ_p) in (4.61) and (4.62) is replaced by (y/τ_p), due to the spatial distribution difference between concentrated and distributed coils. Also s_N is to replace d_w where a wire locates, because a coil as integrity is taken in this model. Therefore (4.60) is finally written as

$$ N(\beta) = \frac{2N_c}{\pi}\sum_{\nu}\frac{{}^{\nu}\xi_N{}^{\nu}\xi_S}{\nu}\cos\left(\frac{\nu}{p}\beta\right), \quad \nu/p = 1, 2, 3\ldots \tag{4.63} $$

And the harmonic form along the complete circumference β can be written as

$$ {}^{\nu}N(\beta) = \frac{N_c}{\pi}\frac{{}^{\nu}\xi_N{}^{\nu}\xi_S}{\nu}\cos\left(\frac{\nu}{p}\beta\right), \quad \nu/p = \pm 1, \pm 2, \pm 3\ldots \tag{4.64} $$

4.2.4 Model of Coil Distribution

In this section, spatial distribution of coils is further deduced. First, both concentrated coil models based on the studied 3-phase SRM in Fig. 4.4 in Sect. 4.2.2 (Model 1) and 4.2.3 (Model 2) are investigated, followed by the distributed coil models in Sect. 4.2.4. Now coils in one phase are numbered with ρ and phases are numbered with k. According to (4.40) and (4.45), the winding function of the first coil ($\rho = 1$) in the first phase ($k = 1$) of Model 1 considers spatial distribution in complex form and now is written as

$$ {}^{\nu}\underline{N}_{1,1}(\beta) = \frac{S}{\nu\pi}{}^{\nu}\xi_{NS}{}^{\nu}\xi_Z e^{-j\frac{\nu}{p}(\beta-\beta_S)} e^{j\frac{\nu}{p}\left(-(S-1)\frac{\pi d_w}{2\tau_{p,\gamma}}\right)}, \quad \nu/p = \pm 1, \pm 2, \pm 3\ldots \tag{4.65} $$

In analogy, the complex form in Model 2 is deduced by neglecting the potential β_S, as well as the distribution of layers, which is alternatively written as

$$ {}^{\nu}\underline{N}_{1,1}(\beta) = \frac{S}{\nu\pi}{}^{\nu}\xi_{NS}{}^{\nu}\xi_Z e^{-j\frac{\nu}{p}\beta}, \quad \nu/p = \pm 1, \pm 2, \pm 3\ldots \tag{4.66} $$

Figure 4.13 shows layout of all layers ($S = 2$) of all coils in one phase. The parameters γ, ρ and k stand for ordinal number of layers, poles and phases respectively, where $\gamma = 1,2,3,\ldots, S$, $\rho = 1,2,3, \ldots, 2p$, and $k = 1,2,3,\ldots N_{ph}$, where N_{ph} is

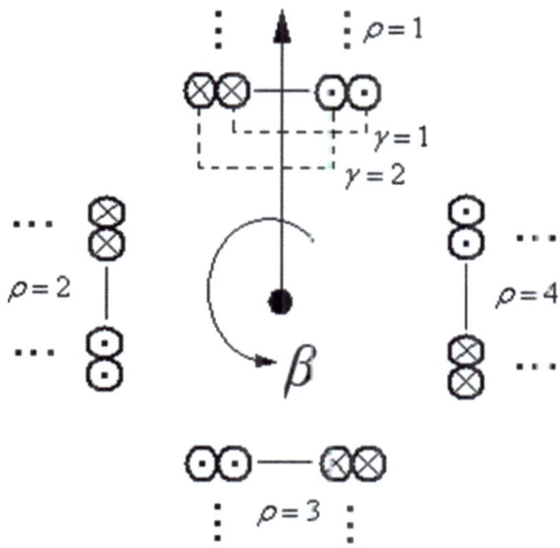

Fig. 4.13 Spatial distribution of all coils

the number of phases. It is stressed that due to SRM topology, one tooth concentrated coil is deemed as one pole. The starting point is selected as vertical upright and Model 1 is analyzed first, with the following couple of assumptions apply,

1. The angular expression of ρ-th coil in one phase is $(\rho-1)p(\pi/2) = \pi(\rho-1)$, with additional part $\pi(\rho-1)$ to account for reversal of DC phase current excitation between a couple of poles that has a mechanical angle of 180°, according to SRM operation principle.
2. The mechanical angular shift of a couple of adjacent stator teeth is $2\pi/4N_{ph} = \pi/2N_{ph}$, where $4N_{ph}$ is the total number of teeth. So altogether the angular shift of k-th phase is expressed as $(k-1)p\pi/2N_{ph}$.

For Model 1, winding function harmonics by ρ-th pole in k-th phase in complex form can be written as

$$^{v}\underline{N}\rho,k(\beta) = \frac{S}{\pi v}{}^{v}\xi_{NS}{}^{v}\xi_{Z}e^{j\frac{v}{p}\left(-(S-1)\frac{\pi d_w}{2\tau_{p,\gamma}}\right)}e^{-j\frac{v}{p}\left(\beta-\beta_S-(\rho-1)\pi-\frac{(k-1)p\pi}{2N_{ph}}\right)}e^{-j(\rho-1)\pi}$$
$$v/p = \pm1,\ \pm2,\ \pm3\ldots \quad (4.67)$$

To sum up all coils, (4.67) is further written as

$$
{}^{v}\underline{N}(\beta) = \frac{S}{\pi v}{}^{v}\xi_{NS}{}^{v}\xi_{Z}e^{j\frac{v}{p}\left(-(S-1)\frac{\pi d_w}{2\tau_{p,\gamma}}\right)}\sum_{k=1}^{N_{ph}}\sum_{\rho=1}^{2p}e^{j\left(-\frac{v}{p}\beta+\frac{v}{p}\beta_S+\frac{v}{p}(\rho-1)\pi+\frac{v}{p}(k-1)\frac{p\pi}{2N_{ph}}-(\rho-1)\pi\right)}
$$
$$
v/p = \pm 1,\ \pm 2,\ \pm 3\ldots \tag{4.68}
$$

By sort of harmonics, it is further written as

$$
{}^{v}\underline{N}(\beta) = \frac{S^{v}\xi_{NS}{}^{v}\xi_{Z}}{\pi v}e^{j\left(-\frac{v}{p}\left(\beta-\beta_S+(S-1)\frac{\pi d_w}{2\tau_{p,\gamma}}\right)\right)}\sum_{\rho=1}^{2p}e^{j(\rho-1)\pi\left(\frac{v}{p}-1\right)}\sum_{k=1}^{N_{ph}}e^{j(k-1)\frac{\pi v}{2N_{ph}}} \tag{4.69}
$$
$$
v/p = \pm 1,\ \pm 2,\ \pm 3\ldots
$$

As to Model 2, the distribution factor does not exist and the slot opening and short pitch factors are separately written. Also the number of layers S and corresponding shifts are not needed. In analogy, the sum of all coils in all phases leads to

$$
{}^{v}\underline{N}(\beta) = \frac{{}^{v}\xi_{N}{}^{v}\xi_{S}}{\pi v}e^{j\left(-\frac{v}{p}\beta\right)}\sum_{\rho=1}^{2p}e^{j(\rho-1)\pi\left(\frac{v}{p}-1\right)}\sum_{k=1}^{N_{ph}}e^{j(k-1)\frac{\pi v}{2N_{ph}}},
$$
$$
v/p = \pm 1,\ \pm 2,\ \pm 3\ldots \tag{4.70}
$$

Then in this section for the distributed coils, a typical topology is shown in Fig. 4.14, where $N_{ph} = 3$ is the number of phases, $q = 3$ is the width per pole per phase measured in number of slots, ε is the step reduction measured in number of slots, τ_N is slot pitch, s_N is the slot opening and y is coil width. The step reduction ε is used as a means of reducing slot pitch, and the relation with y is then written as

$$
y = \left(N_{ph}q - \varepsilon\right)\tau_N \tag{4.71}
$$

In analogy, κ, ρ and k stand for ordinal number of slots per pole, number of poles per phase and number of phases respectively, where $\kappa = 1,2,3\ldots, q$, $\rho = 1,2,3,\ldots,2p$, and $k = 1,2,3,\ldots N_{ph}$. Based on (4.64), the v-th order harmonic of winding function by one coil of N_p turns in κ-th slot of ρ-th pole of k-th phase in complex form is written as

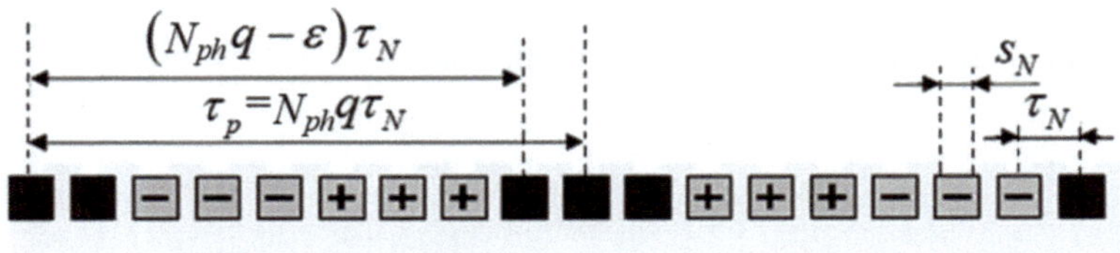

Fig. 4.14 Typical distributed winding topology of an AC machine

$$ {}^{\nu}\underline{N}_{\kappa,\rho,k}(\beta) = \frac{N_c}{\pi} \frac{{}^{\nu}\xi_N {}^{\nu}\xi_S}{\nu} e^{-j\frac{\nu}{p}\beta}, \quad \nu/p = \pm 1, \pm 2, \pm 3 \ldots \tag{4.72} $$

Different from concentrated coil models, by introducing the parameter ε, distribution factor for ν-th harmonic can be now written as follows

$$ \begin{aligned} {}^{\nu}\xi_Z &= \sin\left(\frac{\nu}{p}\frac{y}{\tau_p}\frac{\pi}{2}\right) = \sin\left(\frac{\nu}{p}\frac{(N_{ph}q-\varepsilon)\tau_N}{\tau_p}\frac{\pi}{2}\right) \\ &= \sin\left(\frac{\nu}{p}\frac{(N_{ph}q-\varepsilon)\tau_N}{N_{ph}q\tau_N}\frac{\pi}{2}\right) = \sin\left(\frac{\nu}{p}\frac{N_{ph}q-\varepsilon}{N_{ph}q}\frac{\pi}{2}\right) \\ &= \sin\left(\frac{\nu}{p}\left(1-\frac{\varepsilon}{N_{ph}q}\right)\frac{\pi}{2}\right) \end{aligned} \tag{4.73} $$

Now considering distribution factor, the ν-th order harmonic of winding function by one coil in κ-th slot of ρ-th pole of k-th phase in complex form is written as

$$ {}^{\nu}\underline{N}_{\kappa,\rho,k}(\beta) = \frac{N_c}{\pi}\frac{{}^{\nu}\xi_N {}^{\nu}\xi_S}{\nu} e^{-j\frac{\nu}{p}\left(\beta-(\kappa-1)\frac{\pi}{N_{ph}q}-(\rho-1)\pi-(k-1)\frac{2\pi}{N_{ph}}\right)} e^{-j(\rho-1)\pi}, \\ \nu/p = \pm 1, \pm 2, \pm 3 \ldots \tag{4.74} $$

In analogy to sum all coils, the following assumptions apply:

1. The κ-th coil in a pole has circumferential shift of $x = (\kappa-1)\tau_N$ from the first one. Converted into electrical angle, the shift is alternatively written as $\beta = (\kappa-1)\tau_N(\pi/\tau_p)$. With the expression $\tau_p = N_{ph}q\tau_N$ in Fig. 4.14, it is alternatively as $\beta = (\kappa-1)\pi/(N_{ph}q)$.
2. The ρ-th pole in one phase has circumferential shift in length by $x = (\rho-1)\,N_{ph}q\tau_N$ from the first one. In electrical angle the shift is alternatively written as $\beta = (\rho-1)\pi$.
3. The current excitation is alternating and therefore there must be a direction reversal for every couple of poles. The factor -1, in complex form $e^{-j(\rho-1)\pi}$ is applied.
4. The k-th phase has circumferential shift in length by $x = (k-1)2\tau_p/N_{ph}$ from the first one. In electrical angle the shift is written as $\beta = (k-1)2\tau_p/N_{ph}\,(\pi/\tau_p) = (k-1)\,2\pi/N_{ph}$.

Now to sum up all coils, (4.74) is further written as

$$
{}^{v}\underline{N}(\beta) = \sum_{k=1}^{N_{ph}} \sum_{\rho=1}^{2p} \sum_{\kappa=1}^{q} \frac{N_c}{\pi} \frac{{}^{v}\xi_N {}^{v}\xi_S}{v} e^{j\left(-\frac{v}{p}\beta + \frac{v}{p}(\kappa-1)\frac{\pi}{N_{ph}q} + \frac{v}{p}(\rho-1)\pi + \frac{v}{p}(k-1)\frac{2\pi}{N_{ph}} - (\rho-1)\pi\right)} \quad (4.75)
$$
$$
v/p = \pm 1, \pm 2, \pm 3 \ldots
$$

By sort of harmonics, it is further written as

$$
\begin{aligned}
{}^{v}\underline{N}(\beta) &= \frac{N_c}{\pi} \frac{{}^{v}\xi_N {}^{v}\xi_S}{v} e^{-j\frac{v}{p}\beta} \sum_{k=1}^{N_{ph}} e^{j\frac{v}{p}(k-1)\frac{2\pi}{N_{ph}}} \sum_{\rho=1}^{2p} e^{j\left(\frac{v}{p}(\rho-1)\pi - (\rho-1)\pi\right)} \sum_{\kappa=1}^{q} e^{j\frac{v}{p}(\kappa-1)\frac{\pi}{N_{ph}q}} \\
&= \frac{N_c}{\pi} \frac{{}^{v}\xi N {}^{v}\xi_S}{v} e^{-j\frac{v}{p}\beta} \sum_{k=1}^{N_{ph}} e^{j\frac{v}{p}(k-1)\frac{2\pi}{N_{ph}}} \sum_{\rho=1}^{2p} e^{j\left(\frac{v}{p}-1\right)(\rho-1)\pi} \sum_{\kappa=1}^{q} e^{j\frac{v}{p}(\kappa-1)\frac{\pi}{N_{ph}q}}, \\
&\qquad v/p = \pm 1, \pm 2, \pm 3 \ldots
\end{aligned} \quad (4.76)
$$

4.3 Modeling of Phase Current

In this section, only the DC excitation for SRMs is modeled, and it is not necessary to model AC excitation. The typical phase current waveform is shown in Fig. 4.15, a trapezoid with peak value *I*, as well as the rate of falling and rising. The ripple caused by chopping of power electronics is neglected. The rates of change of current are respectively identified by A_1-A_2 and A_3-A_4 on the horizontal axis of electrical angle which is represented by ωt, where ω and t are respectively the rotating speed of the mechanical rotor and time. Therefore the term ωt together describes angular shift in mechanical degrees. The vertical axis of level of current excitation locates in the middle of a trapezoid and there is a move of $\pi/4$ for the next pole with one phase. The waveform is further simplified by making the rate of rising and falling identical and therefore the trapezoid is evenly symmetrical, which facilitates Fourier

Fig. 4.15 Model of phase current

transformation by neglecting odd components. The term "the rate of change of phase current" is used to describe the rate, which is defined by A_1, A_2 or A_3, A_4. In general the rate of dropping is much more decisive, as airgap reluctance approaches the minimum when phase current is on the way of dropping down, thus making flux variation more influential. Therefore, a couple of parameters θ_1 and θ_2, meaning the turn-off angle and current extinguishing angle, are defined, which respectively corresponds to A_1 and A_2.

For 3-phase excitation, there is one phase current always remaining 0. Using Fourier transformation, the instantaneous current i in Phase k is described as (See Appendix A.2)

$$i_k = \frac{a_0}{2} + \sum_{u=1}^{\infty} \frac{I(\cos(N_r u\theta_1) - \cos(N_r u\theta_2))}{4\pi(\theta_2 - \theta_1)u^2} \cos\left(u\left(N_r\omega t - (k-1)\frac{2\pi}{N_{ph}}\right)\right) \tag{4.77}$$

where u is the order of harmonics and I is the peak value of current $i(\omega t)$, N_r is the number of rotor poles, meaning that the electrical speed is N_r times of the mechanical rotor speed ω. The current of u-th order in complex form in Phase k can be now written as (where complex form is underlined)

$${}^u i_k = \mathrm{Re}\{{}^u\underline{i}_k\}, \quad {}^u\underline{i}_k = {}^u i e^{ju\left(N_r\omega t-(k-1)\frac{2\pi}{N_{ph}}\right)} + \frac{a_0}{2}, \quad {}^u I = \sqrt{2}\,{}^u i \tag{4.78}$$

$${}^u i = \frac{2I}{\sqrt{2}N_r\pi(\theta_2 - \theta_1)u^2}(\cos(N_r\theta_1 u) - \cos(N_r\theta_2 u)) \tag{4.79}$$

$$u = 1, 2, 3 \ldots; \quad k = 1, 2, \ldots, N_{ph}$$

The current waveform is featured by the rate of change. The turn-off angle θ_1 is set by designers while the current extinguishing angle θ_2 may vary by control scheme. The sketch diagram is shown in Fig. 4.16. Let $k = 1$ in (4.77), the current harmonic of u-th order of the first phase is then written as

$${}^u i_1 = \frac{2I}{N_r\pi(\theta_2 - \theta_1)u^2}(\cos(N_r u\theta_1) - \cos(N_r u\theta_2))\cos(N_r u\omega t) \tag{4.80}$$

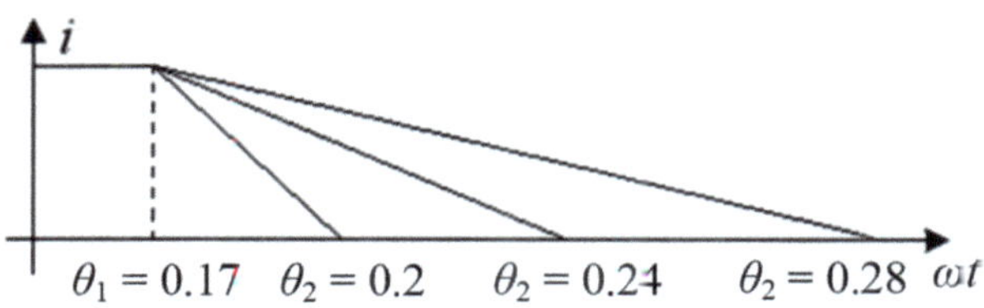

Fig. 4.16 Model of the rates of change of phase current determined by θ_1 and θ_2

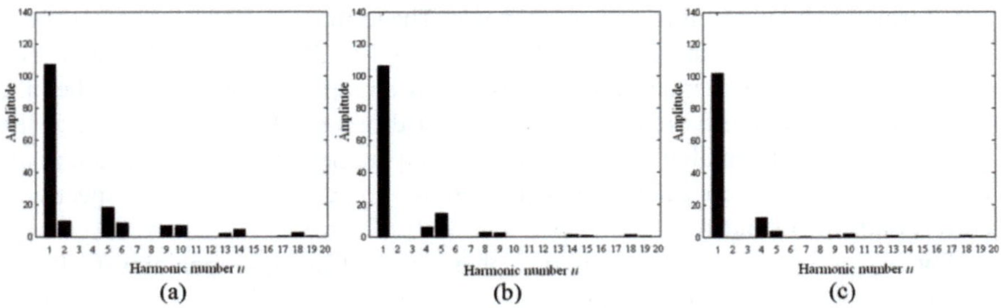

Fig. 4.17 Phase current harmonics by different rates in Fig. 4.16, (**a**) $\theta_2 = 0.2$ rad, (**b**) $\theta_2 = 0.24$ rad, (**c**) $\theta_2 = 0.28$ rad

Let the time $t = 0$, the amplitude of u-th order current harmonic is further written as

$$^{u}I = \frac{2I}{N_r\pi(\theta_2 - \theta_1)u^2}(\cos(N_r u\theta_1) - \cos(N_r u\theta_2)) \tag{4.81}$$

The amplitudes of harmonics with rates of change of current are shown in Fig. 4.17. A lower θ_2 value means a higher rate, which brings amplitude of higher orders more prominent, indicating that more eddy current loss will occur due to harmonics.

4.4 Modeling of MMF Distribution

The MMF distribution is obtained by multiplying winding function and phase current, which has been studied respectively in Sects. 4.2 and 4.3. In this section, winding function models of both concentrated and distributed coils are further developed by adding current excitation to model MMF distribution.

4.4.1 Concentrated Coils

Model 1 with layout of wires inside a coil is developed first. Referring (4.76), the sum of all coils of all phases, and (4.78), the complex form of phase current, the MMF distribution Θ of ν-th harmonic due to u-th current excitation is now written as (where the complex form is underlined)

$$
\begin{aligned}
{}^{u,v}\underline{\Theta}(\beta,t) &= \sqrt{2}\,{}^{u}I\frac{S\,{}^{v}\xi_{NS}\,{}^{v}\xi_{Z}}{\pi v} \quad e^{j\left(N_r u\omega t-\frac{v}{p}\left(\beta-\beta_S+(S-1)\frac{\pi d_w}{2\tau_{p,\gamma}}\right)\right)} \\
&\sum_{\rho=1}^{2p} e^{j(\rho-1)\pi\left(\frac{v}{p}-1\right)} \sum_{k=1}^{N_{ph}} e^{j(k-1)\frac{p\pi}{2N_{ph}}\left(\frac{v}{p}-\frac{4}{p}u\right)}, \\
&v/p = \pm 1,\ \pm 2,\ \pm 3\ldots
\end{aligned} \tag{4.82}
$$

The MMF waveform is a function of circumferential variation β and time t. In analogy, the MMF distribution in Model 2 is written as

$$
\begin{aligned}
{}^{u,v}\underline{\Theta}(\beta,t) &= \sqrt{2}\,{}^{u}I\,\frac{{}^{v}\xi_N\,{}^{v}\xi_S}{\pi v}\,e^{j\left(N_r u\omega t-\frac{v}{p}\beta\right)} \sum_{\rho=1}^{2p} e^{j(\rho-1)\pi\left(\frac{v}{p}-1\right)} \sum_{k=1}^{N_{ph}} e^{j(k-1)\frac{p\pi}{2N_{ph}}\left(\frac{v}{p}-\frac{4}{p}u\right)} \\
v/p &= \pm 1,\ \pm 2,\ \pm 3\ldots
\end{aligned} \tag{4.83}
$$

For both models, the sum of coils and phases is further simplified by sum of unit vectors (See Appendix A.3). Finally, the MMF distribution of Model 1 is written as

$$
\begin{aligned}
{}^{u,v}\underline{\Theta}(\beta,t) &= 2\sqrt{2}\,{}^{u}I\frac{pS_v}{\pi v}\,{}^{v}\xi_L \frac{\sin\left(\left(\frac{v}{p}-2u\right)\frac{p\pi}{4}\right)}{\sin\left(\left(\frac{v}{p}-2u\right)\frac{p\pi}{4N_{ph}}\right)} \\
&e^{j\left(N_r u\omega t-\frac{v}{p}\left(\beta+(S-1)\frac{\pi d_w}{2\tau_{p,\gamma}}-\beta_S-\frac{(N_{ph}-1)}{2}\frac{p\pi}{2N_{ph}}\right)-u\left(N_{ph}-1\right)\frac{p\pi}{2N_{ph}}\right)} \\
&\frac{v}{p} = 2a+1, \quad a = 0,\ \pm 1,\ \pm 2\ldots, \quad u = 1,2,3\ldots
\end{aligned} \tag{4.84}
$$

${}^{v}\xi_L$ includes all factors regarding a layer, ${}^{v}\xi_L = {}^{v}\xi_{NS}\cdot{}^{v}\xi_Z$. Then the winding factor ${}^{v}\xi_W$ is employed, a coefficient combining all the three factors. Here it describes the relationship between the excited MMF, as well as the dimension and position of a single turn. It is written as

$$
{}^{v}\xi_W = \left(\frac{N_p}{S}\right)^{-1}{}^{v}\xi_L = \frac{S}{N_p}{}^{v}\xi_L = \frac{S}{N_p}{}^{v}\xi_{NS}{}^{v}\xi_Z \tag{4.85}
$$

Then (4.84) is finally written as

$$
\begin{aligned}
{}^{u,v}\underline{\Theta}(\beta,t) &= \sqrt{2}^{u}I^{v}\xi_{W}\frac{N_{tph}}{\pi v}\frac{\sin\left(\left(\frac{v}{p}-2u\right)\frac{p\pi}{4}\right)}{\sin\left(\left(\frac{v}{p}-2u\right)\frac{p\pi}{4N_{ph}}\right)} \\
& e^{j\left(N_r u\omega t-\frac{v}{p}\left(\beta+(S-1)\frac{\pi d_W}{2\tau_{p,\gamma}}-\beta_S-\frac{N_{ph}-1}{2}\frac{p\pi}{2N_{ph}}\right)-u\left(N_{ph}-1\right)\frac{p\pi}{2N_{ph}}\right)} \\
& \frac{v}{p} = 2a+1, \quad a=0,\pm 1,\pm 2\ldots, \quad u=1,2,3\ldots
\end{aligned}
\tag{4.85}
$$

where $N_{tph} = 2pN_p$ refers number of turns per phase. In analogy, Model 2 is written based on (4.85), with the following assumptions

$$
\begin{aligned}
{}^{v}\xi_{Z} = 1, \quad S = 1, \quad \sum_{\gamma=1}^{S} e^{j\frac{v}{p}\left(-\pi d_w(\gamma-1)/\tau_{p,\gamma}\right)} = 1, \quad d_w = w_{SS}/2, \\
{}^{v}\xi_{NS} = N_p{}^{v}\xi_{N}{}^{v}\xi_{S}, \quad \tau_{p,\chi} = \tau_p, \quad \beta_S = 0
\end{aligned}
\tag{4.86}
$$

The MMF expression of Model 2 is alternatively written as

$$
\begin{aligned}
{}^{u,v}\underline{\Theta}(\beta,t) &= \sqrt{2}^{u}I^{v}\xi_{W}\frac{N_{tph}}{\pi v}\frac{\sin\left(\left(\frac{v}{p}-2u\right)\frac{p\pi}{4}\right)}{\sin\left(\left(\frac{v}{p}-2u\right)\frac{p\pi}{4N_{ph}}\right)} e^{j\left(N_r u\omega t-\frac{v}{p}\left(\beta-\frac{N_{ph}-1}{2}\frac{p\pi}{2N_{ph}}\right)-u\left(N_{ph}-1\right)\frac{p\pi}{2N_{ph}}\right)} \\
{}^{v}\xi_{W} &= {}^{v}\xi_{N}{}^{v}\xi_{S}, \quad \frac{v}{p} = 2a+1, \quad a=0,\pm 1,\pm 2\ldots, \quad u=1,2,3\ldots
\end{aligned}
\tag{4.87}
$$

4.4.2 Distributed Coils

The MMF distribution of distributed coils is deduced based on (4.76), with a standard k-th phase current in complex form added, which is

$$
\underline{i}_k = Ie^{j(k-1)\frac{2\pi}{N_{ph}}} \tag{4.88}
$$

where I is the effective value of current. Compared with DC excitation in (4.78), the u harmonic or potential $a_0/2$ in is not necessary, then MMF of all poles in all phases are now written as

$$
\begin{aligned}
{}^{\nu}\underline{\Theta}(\beta,t) &= \frac{N_c\sqrt{2}I}{\pi}\,\frac{{}^{\nu}\xi_N{}^{\nu}\xi_S}{\nu}\,e^{j\left(\omega t-\frac{\nu}{p}\beta\right)}\sum_{k=1}^{N_{ph}} e^{j\left(\frac{\nu}{p}(k-1)\frac{2\pi}{N_{ph}}-(k-1)\frac{2\pi}{N_{ph}}\right)}\sum_{\rho=1}^{2p} e^{j\left(\frac{\nu}{p}(\rho-1)\pi-(\rho-1)\pi\right)}\sum_{\kappa=1}^{q} e^{j\frac{\nu}{p}(\kappa-1)\frac{\pi}{N_{ph}q}} \\
&= \frac{N_c I}{\pi}\,\frac{{}^{\nu}\xi_N{}^{\nu}\xi_S}{\nu}\,e^{j\left(\omega t-\frac{\nu}{p}\beta\right)}\sum_{k=1}^{N_{ph}} e^{j\left(\frac{\nu}{p}-1\right)(k-1)\frac{2\pi}{N_{ph}}}\sum_{\rho=1}^{2p} e^{j\left(\frac{\nu}{p}-1\right)(\rho-1)\pi}\sum_{\kappa=1}^{q} e^{j\frac{\nu}{p}(\kappa-1)\frac{\pi}{N_{ph}q}}, \\
&\qquad \nu/p = \pm 1, \pm 2, \pm 3 \ldots
\end{aligned}
\tag{4.89}
$$

Using the following equation, sum of all coils, poles and phases will be performed.

$$
\sum_{\mu=1}^{n} e^{j\mu\gamma} = \frac{\sin\left(n\frac{\gamma}{2}\right)}{\sin\left(\frac{\gamma}{2}\right)}\,e^{j(n+1)\frac{\gamma}{2}}
\tag{4.90}
$$

4.4.2.1 Sum of All Coils Within a Pole

The following is deducted,

$$
\begin{aligned}
&\sum_{\kappa=1}^{q} e^{j\frac{\nu}{p}(\kappa-1)\frac{\pi}{N_{ph}q}} \\
&= e^{-j\frac{\nu}{p}\frac{\pi}{N_{ph}q}}\sum_{\kappa=1}^{q} e^{j\frac{\nu}{p}\kappa\frac{\pi}{N_{ph}q}} \\
&= e^{-j\frac{\nu}{p}\frac{\pi}{mq}}\,\frac{\sin\left(q\frac{1}{2}\frac{\nu}{p}\frac{\pi}{N_{ph}q}\right)}{\sin\left(\frac{1}{2}\frac{\nu}{p}\frac{\pi}{N_{ph}q}\right)}\,e^{j(q+1)\frac{1}{2}\frac{\nu}{p}\frac{\pi}{N_{ph}q}} \\
&= q\,\frac{\sin\left(q\frac{\nu}{p}\frac{\pi}{2N_{ph}q}\right)}{q\sin\left(\frac{\nu}{p}\frac{\pi}{2N_{ph}q}\right)}\,e^{j(q-1)\frac{1}{2}\frac{\nu}{p}\frac{\pi}{N_{ph}q}}, \\
&\frac{\nu}{p}-1 = 2a, \quad a = 0, \pm 1, \pm 2, \ldots
\end{aligned}
\tag{4.91}
$$

Compared with concentrated coils, the distribution factor ${}^{\nu}\xi_Z$ of ν-th order is written as

$$
{}^{v}\xi_Z = \frac{\sin\left(q\frac{v}{p}\frac{\pi}{2N_{ph}q}\right)}{q\sin\left(\frac{v}{p}\frac{\pi}{2N_{ph}q}\right)} \tag{4.92}
$$

Then, sum of all coils in one phase is written as

$$
\begin{aligned}
&\sum_{\kappa=1}^{q} e^{j\frac{v}{p}(\kappa-1)\frac{\pi}{N_{ph}q}} = q\,{}^{v}\xi_Z e^{j(q-1)\frac{1}{2}\frac{v}{p}\frac{\pi}{N_{ph}q}},\\
&\frac{v}{p} - 1 = 2a, \quad a = 0, \pm 1, \pm 2, \ldots
\end{aligned} \tag{4.93}
$$

4.4.2.2 Sum of All Poles Within a Phase

This part is identical to the mathematical deduction of concentrated coils. Refer to Appendix A.3 for details.

4.4.2.3 Sum of All Phases

The following is deducted,

$$
\begin{aligned}
&\sum_{k=1}^{N_{ph}} e^{j\left(\frac{v}{p}-1\right)(k-1)\frac{2\pi}{N_{ph}}} = e^{-j\left(\frac{v}{p}-1\right)\frac{2\pi}{N_{ph}}} \sum_{k=1}^{N_{ph}} e^{j\left(\frac{v}{p}-1\right)k\frac{2\pi}{N_{ph}}}\\
&= e^{-j\left(\frac{v}{p}-1\right)\frac{2\pi}{N_{ph}}} \frac{\sin\left(N_{ph}\frac{1}{2}\left(\frac{v}{p}-1\right)\frac{2\pi}{N_{ph}}\right)}{\sin\left(\frac{1}{2}\left(\frac{v}{p}-1\right)\frac{2\pi}{N_{ph}}\right)} e^{j(N_{ph}+1)\frac{1}{2}\left(\frac{v}{p}-1\right)\frac{2\pi}{N_{ph}}}\\
&= \frac{\sin\left(\left(\frac{v}{p}-1\right)\pi\right)}{\sin\left(\left(\frac{v}{p}-1\right)\frac{\pi}{N_{ph}}\right)} e^{j(N_{ph}-1)\frac{1}{2}\left(\frac{v}{p}-1\right)\frac{2\pi}{N_{ph}}},\\
&\frac{v}{p} - 1 = 2a, \quad a = 0, \pm 1, \pm 2, \ldots
\end{aligned} \tag{4.94}
$$

The term (v/p–1) is always an even number, then

$$
\sin\left(\left(\frac{v}{p}-1\right)\pi\right) = 0 \tag{4.95}
$$

Then, the term $e^{j(N_{ph}-1)\frac{1}{2}\left(\frac{v}{p}-1\right)\frac{2\pi}{N_{ph}}}$ is not null, only under the condition that

$$\sin\left(\left(\frac{v}{p}-1\right)\frac{\pi}{N_{ph}}\right)=0 \tag{4.96}$$

holds true as well. In such occasion, (4.96) is further deduced in the following

$$\begin{aligned}&\left(\frac{v}{p}-1\right)\frac{\pi}{N_{ph}}=b\pi, \quad b=0,\pm 1,\pm 2,\ldots\\ \Rightarrow\quad &\frac{v}{p}-1=bN_{ph}, \quad b=0,\pm 1,\pm 2,\ldots\end{aligned} \tag{4.97}$$

According to (4.94) and (4.97), the followings must be simultaneously satisfied

$$\begin{aligned}&\frac{v}{p}-1=2a \quad \Rightarrow \quad \frac{v}{p}=2a+1, \quad a=0,\pm 1,\pm 2,\ldots\\ &\frac{v}{p}-1=2b \quad \Rightarrow \quad \frac{v}{p}=2b+1, \quad b=0,\pm 1,\pm 2,\ldots\end{aligned} \tag{4.98}$$

(4.98) can be fulfilled if

$$\frac{v}{p}-1=2N_{ph}a \quad \Rightarrow \quad \frac{v}{p}=2N_{ph}a+1, \quad a=0,\pm 1,\pm 2,\ldots \tag{4.99}$$

The sum of all phases is further performed by L'Hospital Law, according to (4.94)

$$\begin{aligned}\sum_{k=1}^{N_{ph}} e^{j\left(\frac{v}{p}-1\right)(k-1)\frac{2\pi}{N_{ph}}} &= \frac{\sin\left(\left(\frac{v}{p}-1\right)\pi\right)}{\sin\left(\left(\frac{v}{p}-1\right)\frac{\pi}{N_{ph}}\right)} e^{j\left(N_{ph}-1\right)\frac{1}{2}\left(\frac{v}{p}-1\right)\frac{2\pi}{N_{ph}}}\\ &= \frac{\pi\cos\left(\left(\frac{v}{p}-1\right)\pi\right)}{\frac{\pi}{N_{ph}}\cos\left(\left(\frac{v}{p}-1\right)\frac{\pi}{N_{ph}}\right)} e^{j\left(N_{ph}-1\right)\frac{1}{2}\left(\frac{v}{p}-1\right)\frac{2\pi}{N_{ph}}}\\ &= N_{ph}\frac{\cos\left(2N_{ph}a\pi\right)}{\cos\left(2N_{ph}a\frac{\pi}{N_{ph}}\right)} e^{j\left(N_{ph}-1\right)\frac{1}{2}2N_{ph}a\frac{2\pi}{N_{ph}}}\\ &= N_{ph}\frac{\cos\left(2N_{ph}a\pi\right)}{\cos\left(2a\pi\right)} e^{j\left(N_{ph}-1\right)2a\pi}, \quad \frac{v}{p}-1=2N_{ph}a, \quad a=0,\pm 1,\pm 2,\ldots\end{aligned} \tag{4.100}$$

For an integer that is a multiplication of 2π, the cosine function equals 1 and therefore, the following applies,

$$
\begin{aligned}
&\sum_{k=1}^{N_{ph}} e^{j\left(\frac{\nu}{p}-1\right)(k-1)\frac{2\pi}{N_{ph}}} = N_{ph}, \\
&\frac{\nu}{p} - 1 = 2N_{ph}a, \quad a = 0, \pm 1, \pm 2, \ldots
\end{aligned}
\tag{4.101}
$$

With sum of all coils in one pole, sum of all poles in one phase and sum of all phases, (4.89) is now written as

$$
\begin{aligned}
{}^{\nu}\underline{\Theta}(\beta, t) &= \frac{N_c\sqrt{2}I}{\pi} \frac{{}^{\nu}\xi_N {}^{\nu}\xi_S}{\nu} e^{j\left(\omega t - \frac{\nu}{p}\beta\right)} N_{ph} 2pq {}^{\nu}\xi_Z e^{j(q-1)\frac{1}{2}\frac{\nu}{p}\frac{\pi}{N_{ph}q}} \\
&= \frac{N_{ph}}{\pi} \frac{2pqN_c {}^{\nu}\xi_N {}^{\nu}\xi_S {}^{\nu}\xi_Z}{\nu} I e^{j\left(\omega t - \frac{\nu}{p}\beta\right)} e^{j(q-1)\frac{1}{2}\frac{\nu}{p}\frac{\pi}{N_{ph}q}}, \\
\frac{\nu}{p} &= 2N_{ph}a + 1, \quad a = 0, \pm 1, \pm 2, \ldots
\end{aligned}
\tag{4.102}
$$

We define the total number of turns in one machine W, and total winding factor ${}^{\nu}\xi_W$, which are

$$
W = 2pqN_c \tag{4.103}
$$

$$
{}^{\nu}\xi_W = {}^{\nu}\xi_N {}^{\nu}\xi_S {}^{\nu}\xi_Z \tag{4.104}
$$

Then (4.102) is further written as

$$
\begin{aligned}
{}^{\nu}\underline{\Theta}(\beta, t) &= \frac{\sqrt{2}N_{ph}}{\pi} \frac{W {}^{\nu}\xi_W}{\nu} I e^{j\left(\omega t - \frac{\nu}{p}\beta\right)} e^{j(q-1)\frac{1}{2}\frac{\nu}{p}\frac{\pi}{N_{ph}q}} \\
&= \frac{\sqrt{2}N_{ph} W {}^{\nu}\xi_W}{\pi\nu} I e^{j\left(\omega t - \frac{\nu}{p}\left(\beta - \frac{1}{2}(q-1)\frac{\pi}{N_{ph}q}\right)\right)}, \\
\frac{\nu}{p} &= 2N_{ph}a + 1, \quad a = 0, \pm 1, \pm 2, \ldots
\end{aligned}
\tag{4.105}
$$

4.5 Simulation and Discussion

In this section, the couple models of concentrated coils for SRMs are simulated and compared. The control parameters are, $f = 50$ Hz, $\omega = 2\pi f$, $\theta_1 = 0.1527$ rad and $\theta_2 = 0.24$ rad respectively, and amplitude of phase current $I = 170$A. The windings parameters are, the number of turns per pole $N_p = 10$, and pole pitch $\tau_p = 0.055$ m. For Model 1, two layers ($S = 2$) are used and additional dimensions are width of teeth $w_{ST} = 0.01$ m, and diameter of wires $d_w = 0.003$ m. For Model 2, width of stator slot is $w_{SS} = 0.0082$ m.

The winding factor harmonics are shown in Fig. 4.18. Compared with distributed coils, harmonics are much higher. There are a couple of dominant Fourier components: ordinal number 3 and 5. Some harmonics become negative with increasing order number ν. By comparison in Model 2, the amplitude of each harmonic is characterized by slot width, but the structure of a coil residing inside is not clear, while in Model 1 amplitudes of harmonics by wire geometry are shown. Negative harmonics are suppressed by a higher diameter value, which means the slot fill factor is preferred higher. The wire diameter $d_w = 0.003$ m is further used.

The developed winding function $\underline{N}(\beta)$ is shown in Fig. 4.19, in which the horizontal axis stands for circumferential direction. Starting from middle axis of a tooth, the 12 trapezoidal waveforms stand for each coil and the amplitude equals to number of turns per pole. The width of a trapezoid is closely related with coil distribution along the air gap direction. In the following, Model 1 is further investigated.

Phase current by Fourier series in (4.81) is added into (4.85) and the MMF harmonics are shown in Fig. 4.20. On the horizontal axis are given fundamental and harmonic waveform numbers, while on vertical axis is the MMF amplitude. The ordinal numbers 1 and 3 are the most obvious and are the working components. Unlike distributed windings, torque ripples are higher. Each harmonic remains constant at different rotor positions 0, $\pi/2$ and π, and therefore time independent torque can be generated. Next, MMF distribution at different positions is simulated. Potential value $a_0/2$ is implemented as part of phase current. Figure 4.21 is used to make a clear current position relationship. One period of a 3-phase current is divided evenly by 6 points from A_1 to A_6. When the current changes to the next phase, for example between A_1 and A_3, the rotor passes an angle of $0.5 \cdot 2\pi/4N_{ph} = \pi/4N_{ph}$, where $4N_{ph}$ refers to number of stator teeth. Between A_1 and A_3 lies A_2 in the middle. The time interval T between simulation points is calculated as

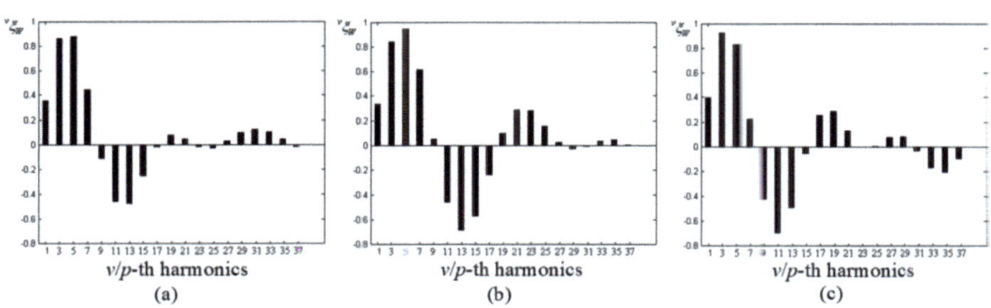

Fig. 4.18 The winding factor harmonics with different wire geometries, (**a**) Model 1, $d_w = 0.003$ m, (**b**) Model 1, $d_w = 0.002$ m, (**c**) Model 2

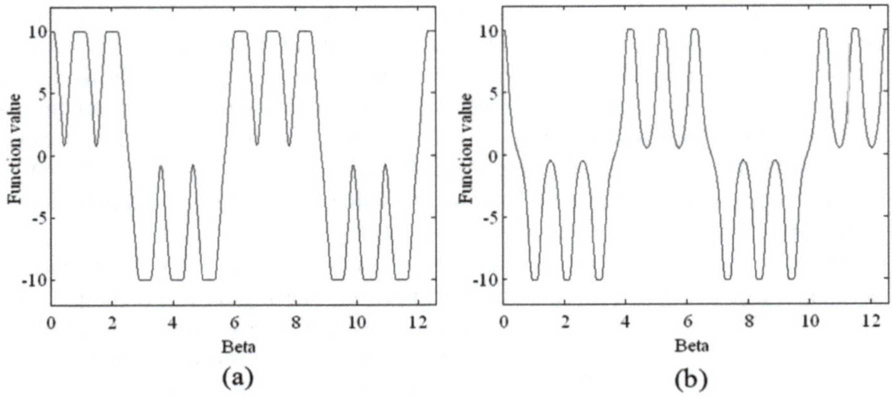

Fig. 4.19 The winding function distribution, (**a**) Model 1, (**b**) Model 2

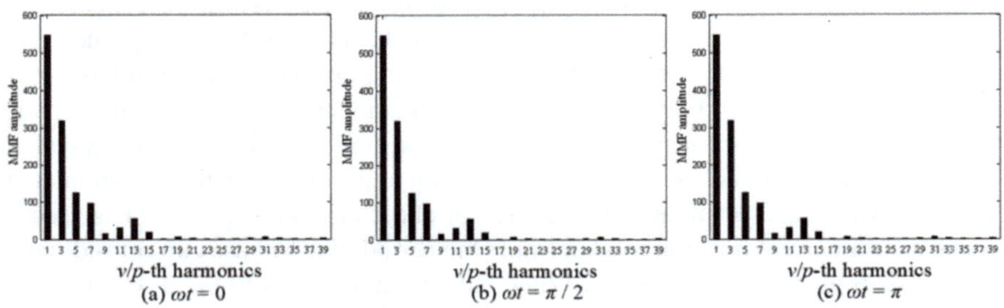

Fig. 4.20 MMF harmonics at different time

$$
\begin{aligned}
T = \theta/\omega &= \frac{1}{2}\left(\frac{1}{2}2\pi/4N_{ph}\right)/\omega = \frac{1}{2}\pi/4N_{ph}\omega \\
&= \frac{1}{2}\pi/(4 \times 3 \times 100 \times 2\pi) = 0.0002083\,\mathrm{s}
\end{aligned}
\tag{4.106}
$$

The curves at different rotor positions are shown in Fig. 4.22. The amplitude changes within the band $\pm N_p I$. The curve goes periodically from (a) to (f), corresponding to the sequence of phase current in Fig. 4.21. Specifically, at the point A_1, A_3 and A_5 only one phase is excited while at A_2, A_4 and A_6 two phases are excited at the same time.

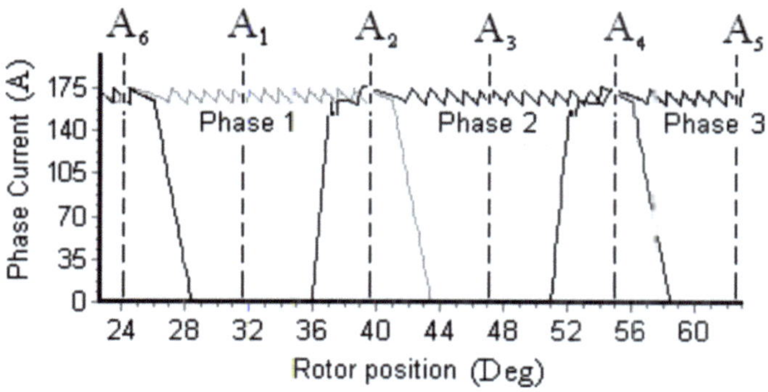

Fig. 4.21 Phase current waveforms and selected simulation points A_1–A_6

4.6 MMF Distribution from Axial Direction

The MMF distribution in axial direction is investigated [6]. Considering end effect, one coil is shown in Fig. 4.23, in which the windings and end windings are defined by lengths L_z and L_{EW_Z} respectively. The arrows indicating direction of current originate from one side of end windings and go vertically into the other side.

The model MMF axial distribution is shown in Fig. 4.24. The axis is set in the middle so that the trapezoidal waveform is of even symmetry. The rate of change of phase current is determined by length of the end windings L_{EW_Z}. With proper extension, such curve is of a 2 l period, where

$$l = L_z + 2L_{EW\_Z} \tag{4.107}$$

Using Fourier transformation, the MMF distribution in axial direction is (where ς is the ordinal number of axial harmonics)

$${}^{\varsigma}f(z) = {}^{\varsigma}a\cos\left(\frac{\pi}{l}\varsigma z\right),\quad \varsigma = 1, 3, 5\ldots \tag{4.108}$$

$${}^{\varsigma}a = \frac{1}{\pi\varsigma}\left(4+\frac{L_z-2l_1}{L_{EW\_Z}}\right)\sin\left(\frac{\pi}{l}l_1\varsigma\right) - \frac{l}{L_{EW\_Z}\pi^2\varsigma^2}\left(\cos\left(\frac{\pi}{l}l_1\varsigma\right)-\cos\left(\frac{\pi}{l}\frac{L_z}{2}\varsigma\right)\right) \tag{4.109}$$

$$l_1 = \frac{L_z}{2} + 2L_{EW\_Z} \tag{4.110}$$

Let $L_z = 127$ mm and $L_{EW\_Z} = 10$ mm for the studied SRM the waveform and harmonics are shown in Fig. 4.25.

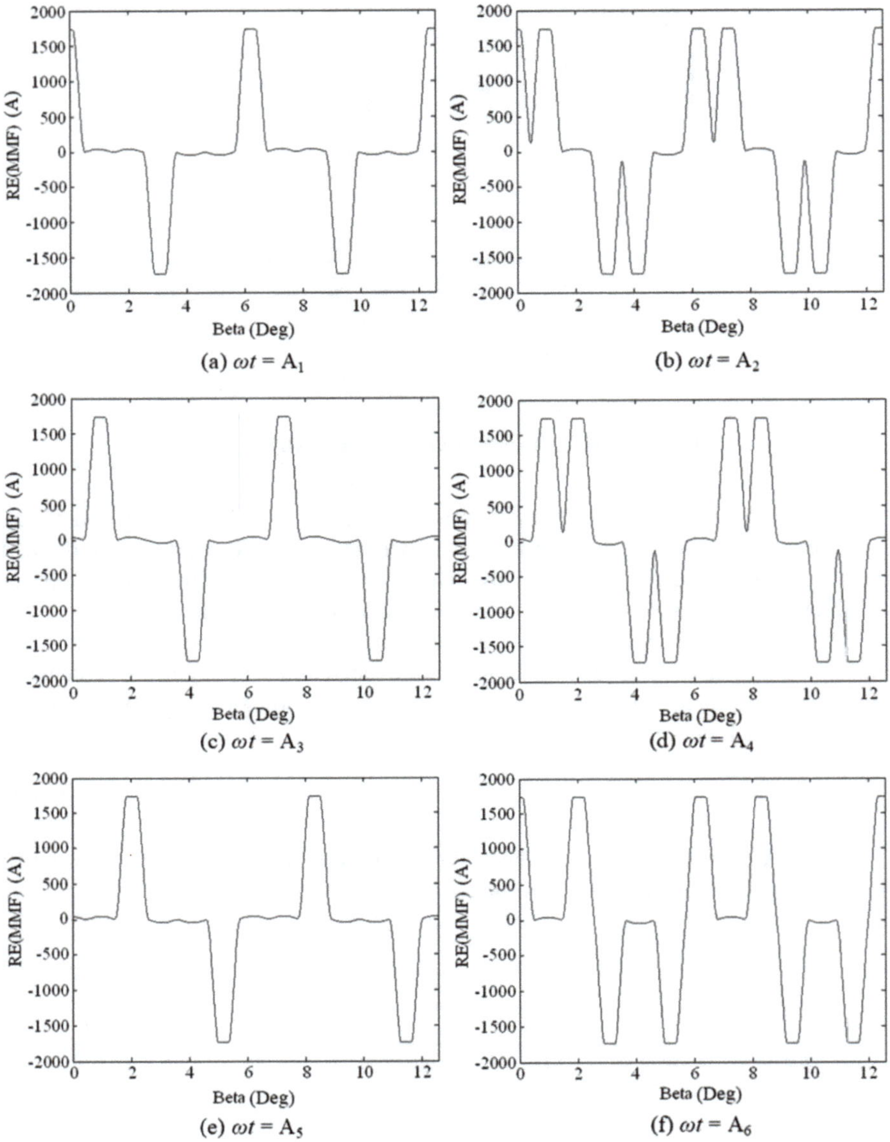

Fig. 4.22 MMF distribution at selected rotor positions

4.7 Application Example

In this section, the developed MMF distribution in 3 dimensions further used to calculate airgap flux density. One simple method exists for electromagnetic analysis. Determination of airgap flux Φ on the following is widely used [7].

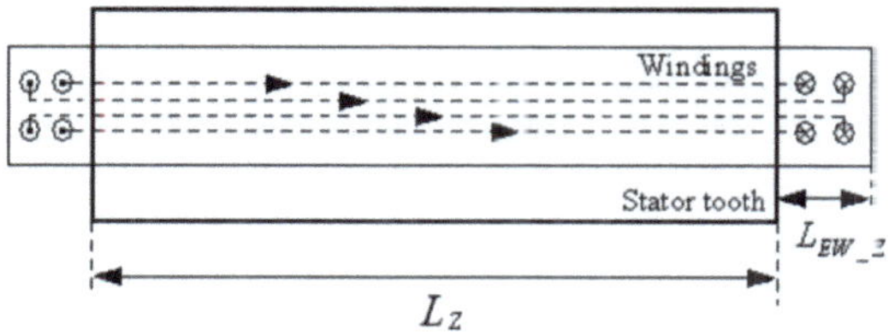

Fig. 4.23 Windings and current flow in axial Z-direction

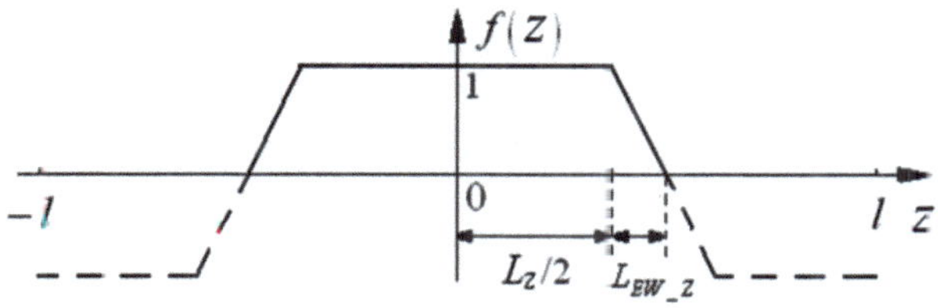

Fig. 4.24 Model of MMF distribution in axial Z-direction

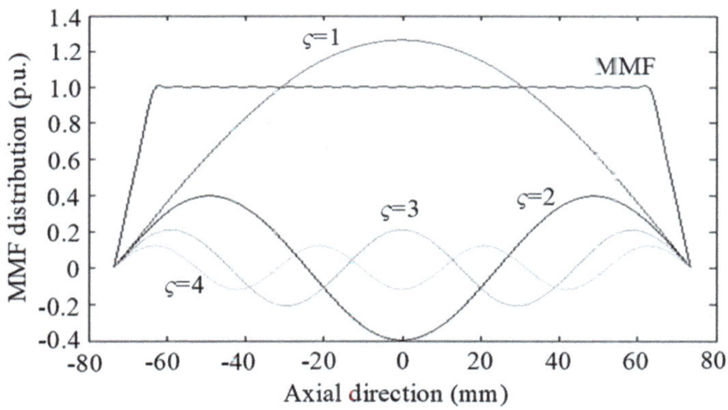

Fig. 4.25 The MMF distribution in axial direction

$$\Phi(\alpha, t) = \Lambda(\alpha, t) \cdot \Theta(\alpha, t) \quad (4.111)$$

where $\Theta(\alpha,t)$ is MMF distribution along airgap circumference α with time variation t, a multiplication of phase current and winding function. $\Lambda(\alpha,t)$ is the distribution of permeance of flux loop, where α is measured in mechanical degrees and $\beta = p\alpha$, where p is the number of pole pairs. $\Lambda(\alpha,t)$ not only includes airgap permeance due to double saliency, but also the magnetic saturation in the circuit.

Then the airgap flux density distribution $B(\alpha,t)$, the characteristic investigated in this section, is obtained by $B = \Phi/A_s$, where A_s is a constant, meaning the area of the cylindrical airgap surface. This simple approximation from (4.111) gives quick estimation of airgap flux density. However, there are shortcomings with regard to either aspect of (4.111). As to MMF, the layout of tooth concentrated wires in slots is not modeled and therefore harmonics due to wire arrangement cannot be analyzed. Accordingly solution is to use the developed Model 1 with layers S. As to airgap permeance, since magnetic path through iron parts is often deem ideal without magnetic saturation, accuracy of this model is limited, especially under high power density. In particular, the saturation is more prominent on SRMs with double saliency. Accordingly solution here is to use the developed discretized circuit network model in Sect. 2.2 of Chap. 2.

Airgap flux density is calculated and FE analysis is used as verification. First to have an overview of the simulated airgap flux density characteristics, the studied SRM is set operating at low speed and conventional fixed angle control mode is used. To facilitate discussion, flux paths at typical rotor positions which correspond to Fig. 2.33 in Chap. 2 is again shown in Fig. 4.26. In (a) only one phase at unaligned position while in (b) a couple of phases, including one close to full alignment with full phase current to be dropped down whereas the other just after the unaligned position where phase current is rising up, are simultaneously excited with current values.

Typical airgap flux density variation along airgap circumference α is calculated via (4.111). The variation at the specific rotor positions corresponding to Fig. 4.26 is calculated and shown in Fig. 4.27. It shows agreement between FE and the proposed method. Specifically in (a), a couple of trapezoids represent the peak of airgap flux by the exciting poles at unaligned position. There is a cavity on top of a trapezoid showing that few fluxes go along middle of teeth into airgap due to large rotor slots. In (b), the density values are the highest at exciting poles of overlap while much lower at non-overlap due to high airgap reluctance. The width of the higher trapezoid indicates the overlap length of the pole pair. Additionally there is the third phase without current and the values are close to zero.

The flux densities by using different layers of a coil at the couple of typical rotor positions are shown in Fig. 4.28. Overall the waveforms are similar for either case. With more layers $S = 3$, the interference fluctuation of airgap flux density close to rotor position value 0 and its neighborhood is reduced. Also, the peak values of trapezoids are enhanced in both figures with more layers.

The corresponding airgap flux density harmonics are shown in Fig. 4.29. The coils with different layers $S = 1,2,3$ are simulated. With more number of layers, the amplitudes of low-order harmonics are enhanced whilst higher orders are mostly suppressed.

4.8 Modeling of Permanent Magnets

The magnetization $\vec{M}$ can be divided into radial M_r and tangential M_α components. In cylindrical coordinate, it is written as [8]

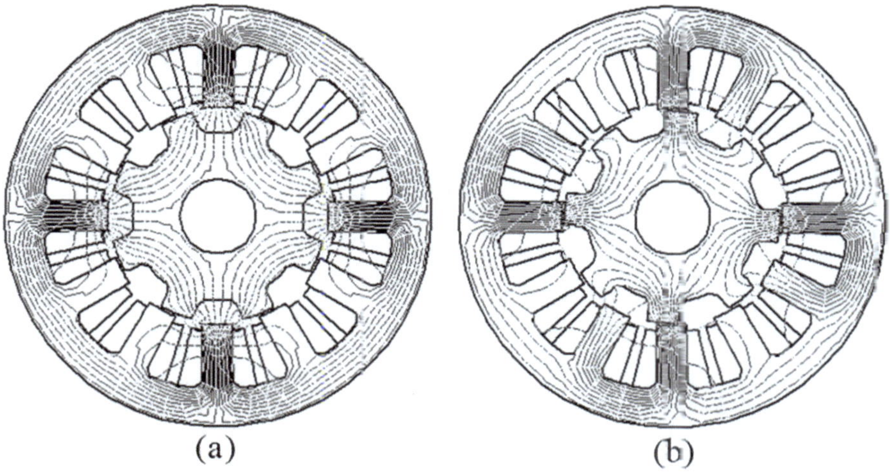

Fig. 4.26 Flux paths overview at typical rotor positions, (**a**) at unaligned position by one single phase excitation, (**b**) at partial aligned position by two phase excitations, showing coupling effect

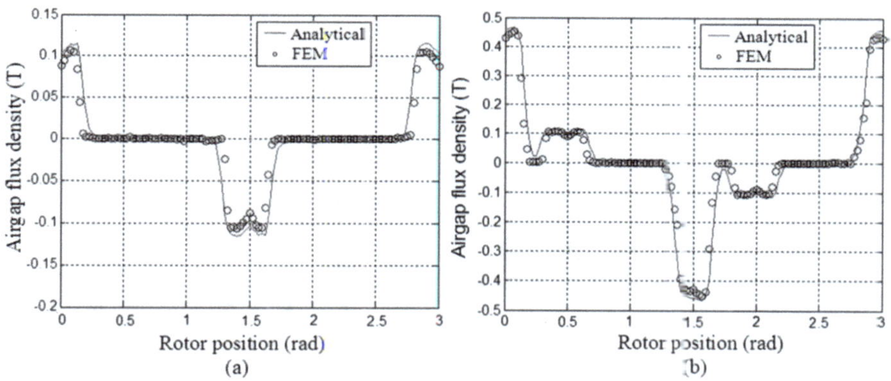

Fig. 4.27 Airgap flux density waveform when $S = 2$, each rotor position corresponds to Fig. 4.26, (**a**) one phase is excited at unaligned position, (**b**) two phases are excited at the same time

$$\vec{M} = M_r \vec{e}_r + M_\alpha \vec{e}_\alpha \tag{4.112}$$

According to magnetization orientation, the MMF variation respectively of radial and diametrical excitations as a function of circumference α in electrical degree is shown in Fig. 4.30, where $\alpha_p = \tau_m/\tau_p$ is the magnet pole-arc to pole-pitch ratio, and B_{rem} is the PM remnant flux density. Over one pole pair, radial and tangential components of the magnetization can be described in [9]. Radial PM excitation of tangential direction M_α always equals zero. For radial PM excitation of radial direction M_r that is non-zero:

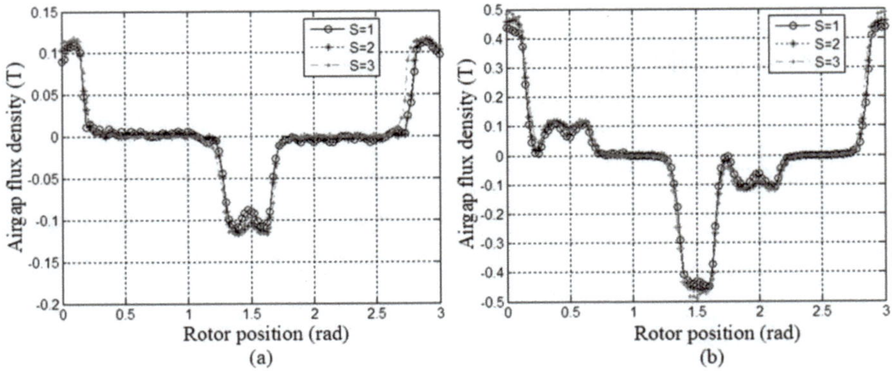

Fig. 4.28 Flux density distribution with different layers of wires $S = 1,2,3$, and each rotor position corresponds to Fig. 4.26, (**a**) one phase is excited at unaligned position, (**b**) two phases are excited at the same time

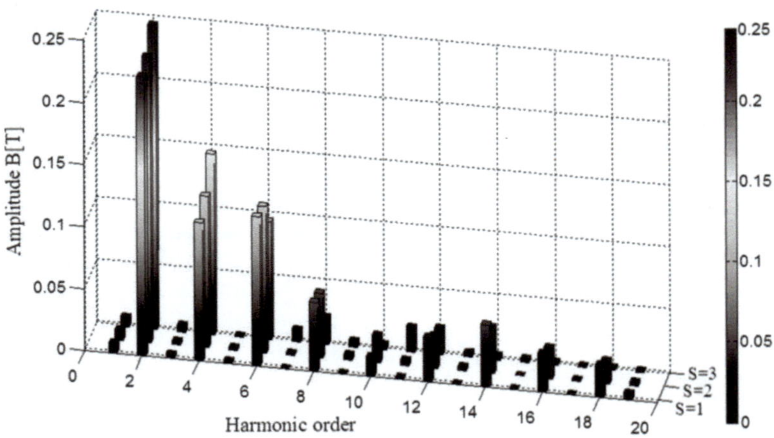

Fig. 4.29 Flux density harmonics with different layers of wires inside a coil

$$M_r = \frac{B_{rem}}{\mu_0}, \quad -\alpha_p\frac{\pi}{2} \le \alpha \le \alpha_p\frac{\pi}{2} \tag{4.113}$$

$$M_r = -\frac{B_{rem}}{\mu_0}, \quad (2-\alpha_p)\frac{\pi}{2} \le \alpha \le (2+\alpha_p)\frac{\pi}{2} \tag{4.114}$$

For diametrical PM excitation of radial direction M_r that is non-zero:

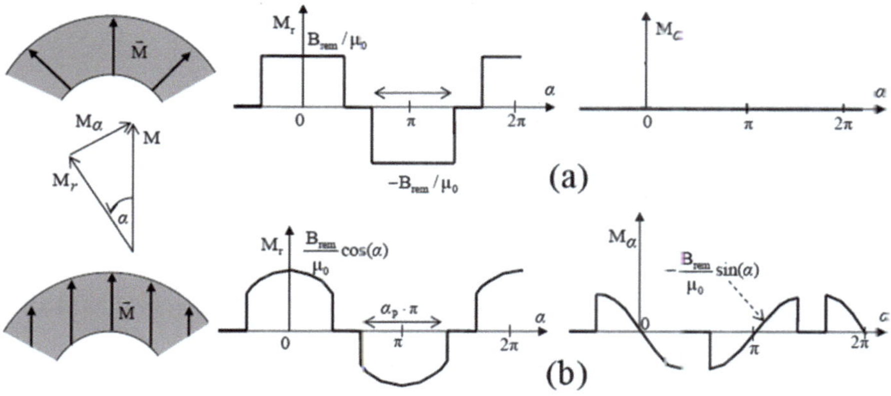

Fig. 4.30 Modeling of magnetization, (**a**) radial, (**b**) diametrical

$$M_r = -\frac{B_{rem}}{\mu_0}\cos(\alpha), \quad -\alpha_p\frac{\pi}{2} \le \alpha \le \alpha_p\frac{\pi}{2}, \quad (2-\alpha_p)\frac{\pi}{2} \le \alpha \le (2+\alpha_p)\frac{\pi}{2} \tag{4.115}$$

For diametrical PM excitation of tangential direction M_α that is non-zero:

$$M_\alpha = -\frac{B_{rem}}{\mu_0}\sin(\alpha), \quad -\alpha_p\frac{\pi}{2} \le \alpha \le \alpha_p\frac{\pi}{2}, \quad (2-\alpha_p)\frac{\pi}{2} \le \alpha \le (2+\alpha_p)\frac{\pi}{2} \tag{4.116}$$

The magnetization is assumed to be uniform throughout the cross section of the magnets. However, it should be noted that the magnetization $\vec{M}$ only exists inside the magnets. Often, magnet pole arcs do not cover the whole rotor surface, and therefore there are air spaces between magnet pole arcs. These air spaces can be treated as magnet arcs without magnetization, as magnets without magnetization are assumed to have the same electromagnetic properties as air. M_r and M_α can be expressed by Fourier series as

$$M_r(\alpha) = \sum_\xi {}^\xi M_r(\alpha), \quad {}^\xi M_r(\alpha) = {}^\xi\widehat{M}_r\cos(\xi\alpha), \quad \xi = 1,3,5\ldots \tag{4.117}$$

$$M_\alpha(\alpha) = \sum_\xi {}^\xi M_\alpha(\alpha), \quad {}^\xi M_\alpha(\alpha) = {}^\xi\widehat{M}_\alpha\cos(\xi\alpha), \quad \xi = 1,3,5\ldots \tag{4.118}$$

Equations (4.117) and (4.118) in complex form is written as

$$\underline{M}_r(\alpha) = \sum_{\xi} {}^{\xi}\underline{M}_r(\alpha), \quad {}^{\xi}\underline{M}_r(\alpha) = {}^{\xi}\widehat{M}_r e^{-j\xi\alpha}, \quad \xi = 1, 3, 5\ldots \tag{4.119}$$

$$\underline{M}_\alpha(\alpha) = \sum_{\xi} {}^{\xi}\underline{M}_\alpha(\alpha), \quad {}^{\xi}\underline{M}_\alpha(\alpha) = j^{\xi}\widehat{M}_\alpha e^{-j\xi\alpha}, \quad \xi = 1, 3, 5\ldots \tag{4.120}$$

For radial magnetization,

$${}^{\xi}\underline{\widehat{M}}_r = 2\frac{B_{rem}}{\mu_0}\alpha_p \frac{\sin\left(\frac{\xi\pi\alpha_p}{2}\right)}{\frac{\xi\pi\alpha_p}{2}} = 2\frac{B_{rem}}{\mu_0}\alpha_p \mathrm{si}\left(\frac{\xi\pi\alpha_p}{2}\right), \quad {}^{\xi}\underline{\widehat{M}}_\alpha = 0 \tag{4.121}$$

For diametrical magnetization,

$${}^{\xi}\underline{\widehat{M}}_r = \frac{B_{rem}}{\mu_0}\alpha_p\left({}^{\xi}A_1 + {}^{\xi}A_2\right), \quad {}^{\xi}\underline{\widehat{M}}_\alpha = \frac{B_{rem}}{\mu_0}\alpha_p\left({}^{\xi}A_1 - {}^{\xi}A_2\right) \tag{4.122}$$

$${}^{\xi}A_1 = \mathrm{si}\left((\xi+1)\alpha_p\frac{\pi}{2}\right), \quad {}^{\xi}A_2 = \mathrm{si}\left((\xi-1)\alpha_p\frac{\pi}{2}\right) \tag{4.123}$$

In axial direction z, it is assumed that the magnetization is uniformly distributed on magnets. In analogy to MMF deduction of armature coils, for a magnet with axial length l_b, the magnetization distribution in axial direction can be presented with a rectangular function $f(z)$, as is shown in Fig. 4.31. Accordingly Fourier discretization is written as

$${}^{\kappa}\mathrm{A} = \frac{1}{l_b}\int_{-l_b}^{l_b} f(z)\cos\left(\frac{\kappa\pi z}{l_b}\right)\mathrm{d}z \quad \kappa = 0, 1, 2\ldots \tag{4.124}$$

$${}^{\kappa}\mathrm{B} = \frac{1}{l_b}\int_{-l_b}^{l_b} f(z)\sin\left(\frac{\kappa\pi z}{l_b}\right)\mathrm{d}z \quad \kappa = 1, 2, 3\ldots \tag{4.125}$$

Based on (4.124) and (4.125), the function $f(z)$ is expressed by Fourier components as

$$f(z) = \frac{{}^{0}\mathrm{A}}{2} + \sum_{\kappa=1}\left({}^{\kappa}\mathrm{A}\cos\left(\frac{\kappa\pi z}{l_b}\right) + {}^{\kappa}\mathrm{B}\sin\left(\frac{\kappa\pi z}{l_b}\right)\right) \tag{4.126}$$

From Fig. 4.31b, Fourier coefficients are written as

$${}^{0}\mathrm{A} = 0, \quad {}^{\kappa}\mathrm{B} = 0, \quad {}^{\kappa}\mathrm{A} = (4/\kappa\pi)\sin(\kappa\pi/2) \tag{4.127}$$

Combining (4.126) and (4.127), the MMF by magnets in axial direction is

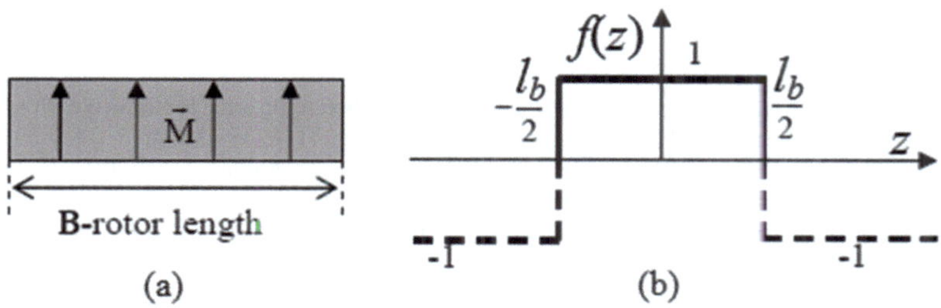

Fig. 4.31 Magnetization in axial direction

$$f(z) = 2\sum_{\kappa=1} \frac{\sin\left(\frac{\kappa\pi}{2}\right)}{\frac{\kappa\pi}{2}} \cos\left(\frac{\kappa\pi}{l_b}z\right) = \sum_{\kappa=1} {}^{\kappa}b \cos\left(\frac{\kappa\pi}{l_b}z\right) \tag{4.128}$$

$${}^{\kappa}b = 2\sin(\kappa\pi/2), \quad \kappa = 1, 3, 5 \ldots$$

4.9 Chapter Summary

Accurate analytical modeling of magneto-motive force (MMF) is an important prerequisite for electromagnetic field analysis. In this chapter, both tooth concentrated and distributed winding topologies are comparatively studied. For the concentrated windings, a couple of models are created whether wire arrangement inside a coil is considered. The first model investigates how turns are arranged within a coil by modeling layers while the second is simplified by considering all turns as integrity. As a result, different winding factors are deduced and discussed. In addition, the machine end part is modeled. Combined with phase current, the MMF distribution in 3 dimensions is simulated. The waveform fits very well with the predicted. This model facilitates winding harmonics analysis, which will be further used to reduce the can loss.

Further by applying a simple magnetic calculation principle, the developed MMF models are applied on a salient SRM to analyze airgap flux density distribution. It is found that the layers of wires inside a coil may have effect on MMF harmonics and airgap flux density is affected as well, which offers possibility to facilitate design of coils. This model can have a fast and accurate magnetic analysis and a concrete view on electromagnetic modeling of electrical machines.

References

1. K. Jaehyuck, R. Krishnan, Novel two-switch-based switched reluctance motor drive for low-cost high-volume applications. IEEE Trans. Ind. Appl. **45**(4), 1241–1248 (2009)
2. L.G. Gallegos, R.J. Krefta, Comparison of full pitch and concentrated coil switched reluctance machines. IEEE Ind. Appl. Conf. **1**, 213–218 (2003)
3. R. Hamdy, J. Fletcher, B.W. Williams, Bidirectional starting of a symmetrical two-phase switched reluctance machine. IEEE Trans. Energy Convers. **15**(1), 211–217 (2000)
4. E.M. Hall, S.S. Ramamurthy, J.C. Balda, Analysis, dimensional sizing and configuration comparison of switched reluctance motors operating under multiphase excitation. IEEE Trans. Energy Convers. **17**(3), 325–311 (2002)
5. Q. Yu, C. Laudensack, D. Gerling, Analysis and loss reduction of a canned switched reluctance drive from the windings perspective, IEEE Power Electronics and Motion Control Conference, ECCE Asia 2012, Harbin, China
6. A. Schram, Redundanzepte fuer Gechaltete Reluktanzantriebe, PhD Dissertation, University of Bundeswehr Muenchen, Munich, Germany, 2006 (in German)
7. P. Frauman, A. Burakov, A. Arkkio, Effects of slot harmonics on the unbalanced magnetic pull in an induction motor with an eccentric rotor. IEEE Trans. Magn. **43**(8), 3441–3444 (2007)
8. D. Gerling, Three dimensional analytical calculation of the permanent magnet motor in cylindrical coordinates, Technical report, Professur fuer Antrienbstechnik und Automation, University of Bumdeswehr Muenchen, Munich, Germany, 2003
9. Z.Q. Zhu, D. Howe, C.C. Chan, Improved analytical model for predicting the magnetic field distribution brussless permanent magnet machines. IEEE Trans. Magn. **38**(1), 229 (2002)

Chapter 5
An Analytical Model of Concentric Layer Structure for Canned Machines, Part II: Magnetic Field

5.1 Model Introduction

The can loss is the main constraint to high efficient and high power density operation of canned machines and most analyses are FE based or with resort to empirical equations. Analytical study of the can effect is necessary, which leads to detailed electromagnetic analysis, that plays the central role in predicting output characteristics. Besides the numerical methods [1, 2], analyses fall into semi-analytical [3, 4] or analytical methods. As to the analytical, a couple of alternatives are predominantly developed. One is the Maxwell's theorem [5–11]. Specifically in [5, 6], the machine is divided into sub-domains (iron cores, airgap, slots and magnets). Our previous work has been applied on an induction, permanent magnet or switched reluctance machine [8, 12, 13]. Magnetic vector potential of each domain is calculated based on magneto-motive force (MMF) distribution as prerequisite. For simplicity a smooth airgap of constant radial length is assumed. By applying Carter's factor [14] or airgap permeance function [8], the airgap is radially enlarged to account for slotting effect. Another alternative focuses on airgap flux density by studying respectively airgap permeance function and MMF distribution [9]. However when cans are concerned, both of these methods may cause numerical deviation.

In this chapter, an analytical model of concentric layer structure is deduced to analyze the electro- magnetic effect due to the use of can shields. Fundamental theories of this chapter, in part, are from [15, 16]. The rotor is divided into concentric cylindrical layers and each is analyzed based on Maxwell equations in 3 dimensions For a canned induction machine, the model structure is shown in Fig. 5.1. In Fig. 5.1a, each part of the machine, including stator, stator can, rotor can, airgap, rotor and shaft, is seen as a concentric hollow cylinder. Each cylinder is geometrically defined with radius r_0–r_6. r_0–r_3 describes the space of air where both layers of cans are accommodated. The stator can layer adheres to stator inner bore whereas the rotor can layer surrounds rotor outer bore. Therefore, the actual airgap is r_1–r_2

Q. Yu et al., *Analysis and Mathematical Models of Canned Electrical Machine Drives*, https://doi.org/10.1007/978-981-13-2745-2_5

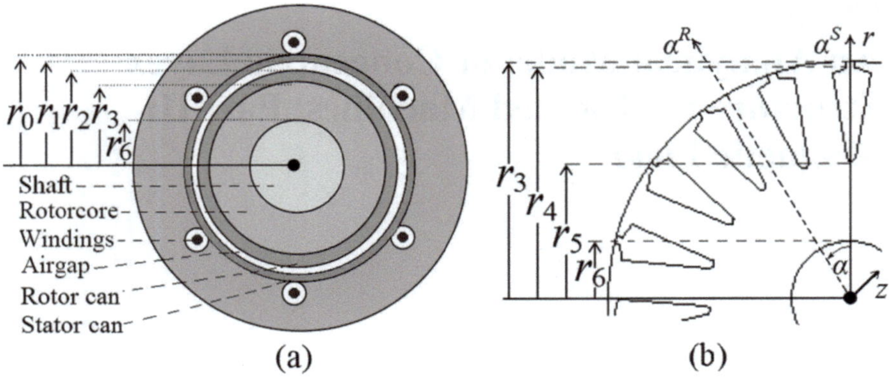

Fig. 5.1 The proposed concentric multiple-layer structure model of a canned induction machine, (**a**) cylindrical components and settings, (**b**) rotor details and the coordinate definition

between the cans. If electrical resistivity of the cans is set 0, the machine is reduced to an ordinary prototype by merging r_0–r_3 as airgap. Then each layer is numbered, stating from 1. Specifically, the stator can layer is the first layer $i = 1$ with thickness r_0–r_1 while the shaft layer is $i = 7$ with thickness r_6–r_7 ($r_7 = 0$, the coordinate origin point). For each layer i, constant values of electrical resistance γ_i and magnetic permeance μ_i are respectively designated. Figure 5.1b shows the solid rotor structure. The rotor outside r_3, top and bottom of rotor slots r_4–r_5 and shaft r_6 are defined respectively. For layers within r_3-r_5 due to rotor inductors, γ and μ are not a constant in radial direction. Lamination steel and rotor inductors are together made up for layers $i = 4$ and $i = 5$ and thus corresponding γ_i and μ_i values are obtained according to the occupied percentage of each material. Also the cylindrical coordinate r-α-z is defined. The upright axis is the starting point and the solid rotor is assumed counterclockwise in terms of rotor position α as positive. Superscripts "*S*" or "*R*" of α indicate the coordinate is whether stator or rotor fixed. If the axis rotates simultaneously with rotor, easier solution can be obtained.

Alternatively for a canned permanent magnet synchronous machine (PMSM), this model is now presented in Fig. 5.2a. In analogy, each radius is labelled from r_0–r_5. r_0 and r_2 are respectively the inner stator and outer rotor bore radius and r_0–r_2 is the airgap length including stator can thickness. The rotor dimension is shown in Fig. 5.2b that the rotor can thickness is r_2–r_3 and there is a layer with radial length r_3–r_4 for permanent magnets (PMs). Each layer is numbered as follows. The stator can is the first layer $i = 1$, and the number gradually increases towards the next inner one till the shaft layer $i = 6$.

For a canned switched reluctance machine, detailed model of rotor and airgap is shown in Fig. 5.3. Each part is considered as a smooth cylindrical layer, including the stator can, air gap, rotor can, rotor and shaft. r_0 is the stator bore radius, r_1 is the outer radius of the air gap, r_2 is the outer radius of the rotor can, r_3 is the outer radius of the rotor and r_4 is the radius of the shaft. Meanwhile, each layer is numbered. The layer of the stator can is the first layer $i = 1$ while the innermost layer (shaft) is $i = 5$.

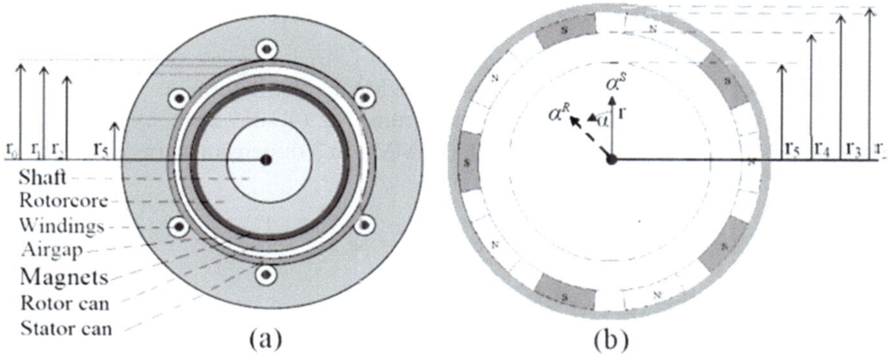

Fig. 5.2 The concentric multiple-layer structure model of a canned PM machine, (**a**) overview, (**b**) rotor details with PMs and the rotor can, with cylindrical coordinate defined

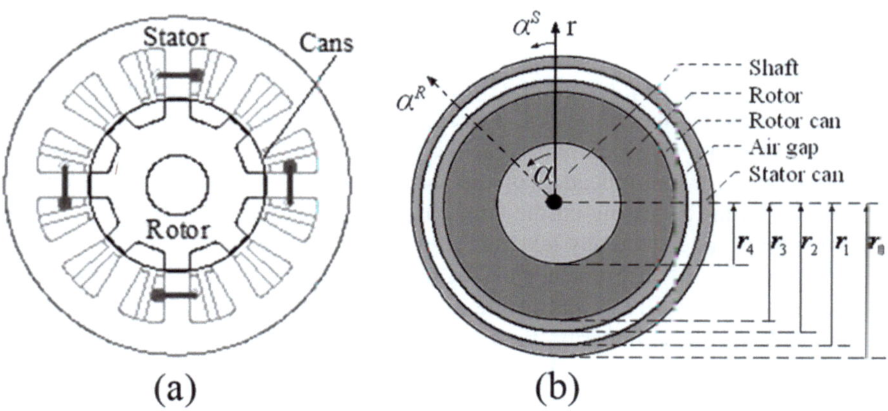

Fig. 5.3 The concentric multiple-layer structure model of a canned switched reluctance machine, (**a**) overview, (**b**) rotor details, the stator can and coordinate definition

Note that due to the saliency of both rotor and stator, the airgap radial length varies with rotor position.

Chapter organization is as follows. In Sect. 5.2, mathematical model of multiple-layer structure is deduced. Starting from the fundamental theory of magnetic field, magnetic vector potential of the first layer is deduced, followed by further layers, in which coordinate transformation and C_i constants are intensively described. Therefore, magnetic field of each layer due to each phase current harmonic is obtained. Further in Sect. 5.3, magnetic field due to current potential of DC excitation is deduced. In Sect. 5.4 as supplement, the rotor position dependent airgap is modeled in case of saliency or severe magnetic saturation. Further in Sect. 5.5, features such as flux density, flux linkage, iron loss, can loss and torque are deducted. Finally in Sect. 5.6, typical canned machines, including the induction machine, PM machine and switched reluctance machine are respectively analyzed, with particular attention on loss, torque and flux due to the use of cans.

5.2 The Model of Concentric Layer Structure

The model of concentric layers starts from MMF distribution in Chap. 4. For distributed coils with AC excitation, according to (4.105) with axial direction considered (4.108), (4.109) and (4.110), the MMF in 3 dimensions now is written as

$$
\begin{aligned}
{}^{\varsigma,\nu}\underline{\Theta}(\alpha,z,t) &= \frac{\sqrt{2}N_{ph}W^{\nu}\xi_W}{\pi\nu} I^{\varsigma}a\cos\left(\varsigma\frac{\pi}{L_z}z\right)e^{j\left(\omega t-\nu_p\left(\alpha-\frac{1}{2}(q-1)\frac{\pi}{N_{ph}q}\right)\right)} \\
&= \frac{\sqrt{2}N_{ph}W^{\nu}\xi_W}{\pi\nu} I^{\varsigma}a\cos\left(\varsigma\frac{\pi}{L_z}z\right)e^{j\left(\omega t-\nu_p(\alpha+\beta_1)+t_{u1}\right)} \\
\nu_p = \nu/p,\quad \nu &= p\left(2N_{ph}a+1\right),\quad a=0,\pm 1,\pm 2\ldots,\quad \varsigma=1,3,5\ldots
\end{aligned} \tag{5.1}
$$

where complex form is underlined. α is the electrical angle, β_1 is an angular shifting constant, t_{u1} is the time shifting constant depending on u (the phase current harmonic if applicable) and it equals to 0 for this application. z is axial variation and t is the time. Superscripts ν and ς indicate the circumferential and axial harmonics respectively. $0 \leq \alpha \leq 2\pi$, $-L_z/2 \leq z \leq L_z/2$, L_z is the axial length, p is the number of pole pairs, N_{ph} is the number of phases, W is the number of turns of the machine, I is the effective value of phase current, ω is the rotating speed. ${}^{\nu}\xi_W$ represents the total winding factor, a multiplication of the slot opening, short pitch and distribution factor. ${}^{\varsigma}a$ and the following cosine term $\cos(\varsigma\pi z/L_z)$ represents MMF along shaft z. End windings are collectively taken in ${}^{\varsigma}a$ in (4.109) in Sect. 4.6.

Alternatively for distributed coils with DC excitation, according to (4.85) of Model 1 with layout of layers inside a coil or simplified Model 2 that takes all turns inside a coil as integrity, the axial distribution is taken as well by (4.108), (4.109) and (4.110) and the MMF in three dimensions now is written as

$$
\begin{aligned}
{}^{u,\varsigma,\nu}\underline{\Theta}(\alpha,z,t) &= \sqrt{2}\,{}^{u}I\,{}^{\nu}\xi_W\frac{W}{\pi\nu}{}^{\varsigma}a\cos\left(\varsigma\frac{\pi}{L_z}z\right)\frac{\sin\left((\nu_p-2u)\frac{p\pi}{4}\right)}{\sin\left((\nu_p-2u)\frac{p\pi}{4N_{ph}}\right)} \\
&\quad e^{j\left(N_r u\omega t-\nu_p\left(\alpha+(S-1)\frac{\pi d_W}{2\tau_{p,\gamma}}-\beta_S-\frac{N_{ph}-1}{2}\frac{p\pi}{2N_{ph}}\right)-u\left(N_{ph}-1\right)\frac{p\pi}{2N_{ph}}\right)} \\
&= \sqrt{2}\,{}^{u}I\,{}^{\nu}\xi_W\frac{W}{\pi\nu}{}^{\varsigma}a\cos\left(\varsigma\frac{\pi}{L_z}z\right)\frac{\sin\left((\nu_p-2u)\frac{p\pi}{4}\right)}{\sin\left((\nu_p-2u)\frac{p\pi}{4N_{ph}}\right)}e^{j\left(N_r u\omega t-\nu_p(\alpha+\beta_2)+t_{u2}\right)} \\
\nu_p &= 2a+1,\quad a=0,\pm 1,\pm 2\ldots,\quad u=1,2,3\ldots
\end{aligned} \tag{5.2}
$$

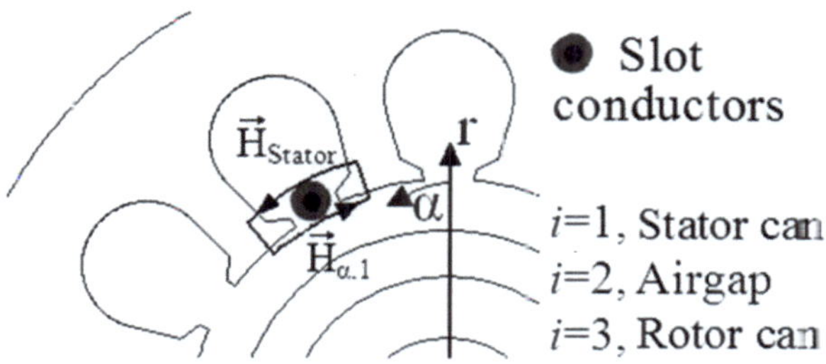

Fig. 5.4 Boundary connection between inner surface of stator bore and outer surface of the first layer

$$
{}^{u,\varsigma,v}\underline{\Theta}(\alpha, z, t) = \sqrt{2}^{u} I^{v} \xi_{W} \frac{W}{\pi v} {}^{\varsigma}a \cos\left(\varsigma \frac{\pi}{L_Z} z\right) \frac{\sin\left(\left(v_p - 2u\right)\frac{p\pi}{4}\right)}{\sin\left(\left(v_p - 2u\right)\frac{p\pi}{4N_{ph}}\right)}
$$

$$
e^{j\left(N_r u\omega t - v_p\left(\alpha - \frac{N_{ph}-1}{2}\frac{p\pi}{2N_{ph}}\right) - u\left(N_{ph}-1\right)\frac{p\pi}{2N_{ph}}\right)} \tag{5.3}
$$

$$
= \sqrt{2}^{u} I^{v} \xi_{W} \frac{W}{\pi v} {}^{\varsigma}a \cos\left(\varsigma \frac{\pi}{L_z} z\right) \frac{\sin\left(\left(v_p - 2u\right)\frac{p\pi}{4}\right)}{\sin\left(\left(v_p - 2u\right)\frac{p\pi}{4N_{ph}}\right)} e^{j\left(N_r u\omega t - v_p(\alpha+\beta_3) + t_{u2}\right)}
$$

$$
v_p = 2a + 1, \quad a = 0, \pm 1, \pm 2 \ldots, \quad u = 1, 2, 3 \ldots
$$

where β_2, β_3 are angular shifting depending on the model and t_{uz} is the time shifting depending on u, and it equals to the term in (5.2) and (5.3) accordingly. Compared with (5.1), another ordinal number u is introduced, which donates the current excitation harmonic. W alternatively means the total number of turns per phase, and N_r is the number of rotor poles that is equivalent to the number of pole pairs for AC machines.

Stator windings and thus MMF by current excitation are transferred to the outer surface of the first layer $i = 1$ (Fig. 5.4). Due that tangential component of the magnetic field strength $\underline{H}$ has to be the same on both inner surface of stator bore and outer surface of the first layer, the magnetic field strength of the first layer $H_{\alpha,1}$ is first deduced. According to the magnetic circuit law $\int \underline{H}_0 \cdot dl = \Theta$, we get (Note that minus sign indicates that magnetic fields are in reverse direction and r is the radius of that layer. Note that the current harmonic u may not apply if the magnetic field excitation does not contain harmonics),

$$
-{}^{u,\varsigma,v}\underline{H}_{\alpha,1} = \frac{1}{r}\frac{\partial}{\partial \alpha} {}^{u,\varsigma,v}\underline{\Theta}(\alpha, z, t) \tag{5.4}
$$

For each layer, the differential equation for the magnetic vector potential $\vec{\underline{A}}$ is (where the arrow indicates vector form),

$$
\Delta \vec{\underline{A}} = \mu\gamma \frac{\partial}{\partial t} \vec{\underline{A}} \tag{5.5}
$$

It is assumed that the stator current only flows in circumferential α and axial z direction, meaning that the eddy current in the rotor and cans is only flowing the paths parallel to stator surface, there does not exist the component $\underline{A}_r$, therefore, (5.5) is simplified as (where $\vec{e}$ is the unit vector)

$$\vec{\underline{A}} = \underline{A}_\alpha \vec{e}_\alpha + \underline{A}_z \vec{e}_z \tag{5.6}$$

Under cylindrical coordinate, (5.5) is further written as

$$\Delta\vec{\underline{A}} = \frac{1}{r}\frac{\partial}{\partial r}\vec{\underline{A}} + \frac{\partial^2}{\partial r^2}\vec{\underline{A}} + \frac{1}{r^2}\frac{\partial^2}{\partial \alpha^2}\vec{\underline{A}} + \frac{\partial^2}{\partial z^2}\vec{\underline{A}} = \mu\gamma\frac{\partial}{\partial t}\vec{\underline{A}} \tag{5.7}$$

Using Maxwell's equations in cylindrical coordinate, the magnetic vector potential $\underline{A}$ of i-th layer in complex form in circumferential α and axial z directions are respectively written as (Note that for AC excitation, the harmonics u does not exist.)

$$\frac{1}{r}\frac{\partial}{\partial r}{}^{u,\varsigma,v}\underline{A}_{\alpha,i} + \frac{\partial^2}{\partial r^2}{}^{u,\varsigma,v}\underline{A}_{\alpha,i} + \frac{1}{r^2}\frac{\partial^2}{\partial \alpha^2}{}^{u,\varsigma,v}\underline{A}_{\alpha,i} + \frac{\partial^2}{\partial z^2}{}^{u,\varsigma,v}\underline{A}_{\alpha,i} = \mu_i\gamma_i\frac{\partial}{\partial t}{}^{u,\varsigma,v}\underline{A}_{\alpha,i} \tag{5.8}$$

$$\frac{1}{r}\frac{\partial}{\partial r}{}^{u,\varsigma,v}\underline{A}_{z,i} + \frac{\partial^2}{\partial r^2}{}^{u,\varsigma,v}\underline{A}_{z,i} + \frac{1}{r^2}\frac{\partial^2}{\partial \alpha^2}{}^{u,\varsigma,v}\underline{A}_{z,i} + \frac{\partial^2}{\partial z^2}{}^{u,\varsigma,v}\underline{A}_{z,i} = \mu_i\gamma_i\frac{\partial}{\partial t}{}^{u,\varsigma,v}\underline{A}_{z,i} \tag{5.9}$$

where A is the magnetic vector potential, the subcript i means the ordinal number of layers. μ_i and γ_i are respectively magnetic permeability and electrical resistivity of i-th layer.

5.2.1 Vector Potential of the First Layer

Equations (5.8) and (5.9) are solved by separation of variables, starting from the first layer $i = 1$. The magnetic field $\underline{H}$ linked with Θ in (5.4) is used. We choose the following setup,

$${}^{u,\varsigma,v}\underline{A}_{z,i} = {}^{u,v}\underline{\Lambda}_i(\alpha){}^{u,\varsigma,v}\underline{R}_i(r){}^{u,\varsigma}\underline{Z}_i(z){}^{u}\underline{T}(t) \tag{5.10}$$

where Λ_i, $\underline{Z}_i$ and $\underline{T}$ are components with regard to the MMF distribution of circumferential, axial and time domain, $\underline{R_i}$ refers to the radius of i-th layer. The 4 components are independent of each other and are respectively written as

$${}^{u,v}\underline{\Lambda}_i(\alpha) = {}^{u,v}\underline{C}_{1,i}e^{-j\frac{v}{p}(\alpha+\beta)} \tag{5.11}$$

$$ {}^{u,\varsigma}\underline{Z}_i(z) = {}^{u,v}\underline{C}_{2,i}{}^{\varsigma}a\cos\left(\varsigma\frac{\pi}{L_z}z\right) \tag{5.12} $$

$$ {}^{u}\underline{T}(t) = e^{j\left({}^{u}\omega t + t_u\right)} \tag{5.13} $$

where $\beta = \beta_1$, β_2, or β_3 and $t_u = t_{u1}$ or t_{u2} are collectively defined and the value depends on the MMF model in (5.1), (5.2), and (5.3), ${}^{u}\omega = u\omega$ is the speed of u-th harmonic. ${}^{u,v}\underline{C}_{1,i}$ and ${}^{u,v}\underline{C}_{2,i}$ are constants. Insert (5.11), (5.12), and (5.13) into (5.10), we get

$$ \begin{aligned} &\frac{1}{r}{}^{u,v}\underline{\Lambda}_i{}^{u,\varsigma,v}\underline{R}'_i{}^{u,\varsigma}\underline{Z}_i{}^{u}\underline{T} + {}^{u,v}\underline{\Lambda}_i{}^{u,\varsigma,v}\underline{R}''_i{}^{u,\varsigma}\underline{Z}_i{}^{u}\underline{T} + \frac{1}{r^2}{}^{u,v}\underline{\Lambda}''_i{}^{u,\varsigma,v}\underline{R}_i{}^{u,\varsigma}\underline{Z}_i{}^{u}\underline{T} \\ &+ {}^{u,v}\underline{\Lambda}_i{}^{u,\varsigma,v}\underline{R}_i{}^{u,\varsigma}\underline{Z}''_i{}^{u}\underline{T} \\ &= \mu_i\gamma_i{}^{u,v}\underline{\Lambda}_i{}^{u,\varsigma,v}\underline{R}_i{}^{u,\varsigma}\underline{Z}_i{}^{u}\underline{T}' \end{aligned} \tag{5.14} $$

Equations (5.11), (5.12), and (5.13) lead to

$$ {}^{u,v}\underline{\Lambda}''_i = \left(-j\frac{v}{p}\right)^2{}^{u,v}\underline{\Lambda}_i = -v_p^2{}^{u,v}\underline{\Lambda}_i \tag{5.15} $$

$$ {}^{u,\varsigma}\underline{Z}''_i = -{}^{u,v}\underline{C}_{2,i}{}^{\varsigma}a\left(\varsigma\frac{\pi}{L_z}\right)^2\cos\left(\varsigma\frac{\pi}{L_z}z\right) = -\left(\varsigma\frac{\pi}{L_z}\right)^2{}^{u,\varsigma}\underline{Z}_i \tag{5.16} $$

$$ {}^{u}\underline{T}' = j^{u}\omega^{u}\underline{T} \tag{5.17} $$

Take (5.15), (5.16) and (5.17) into (5.14), we get

$$ r^2{}^{u,\varsigma,v}\underline{R}''_i + r^{u,\varsigma,v}\underline{R}'_i - \left(\left({}^{u,\varsigma}\underline{k}_i\right)^2r^2 + v_p^2\right){}^{u,\varsigma,v}\underline{R}_i = 0 \tag{5.18} $$

$$ {}^{u,\varsigma}\underline{k}_i = \sqrt{\left(\varsigma\frac{\pi}{L_z}\right)^2 + j^{u}\omega\mu_i\gamma_i} \tag{5.19} $$

which is called the modified Bessel's differential equation of v-th order. The general solution is

$$ {}^{u,\varsigma,v}\underline{R}_i(r) = {}^{u,\varsigma,v}\underline{C}_3\underline{I}_{v_p}\left({}^{u,\varsigma}\underline{k}_i r\right) + {}^{u,\varsigma,v}\underline{C}_4\underline{K}_{v_p}\left({}^{u,\varsigma}\underline{k}\, r\right) \tag{5.20} $$

where $\underline{I}_v$ is the modified Bessel function of first kind and v-th order, $\underline{K}_v$ is the modified Bessel function of second kind and v-th order, ${}^{u,\varsigma,v}\underline{C}_3$ and ${}^{u,\varsigma,v}\underline{C}_4$ are constants. According to (5.20), the solution of the magnetic vector potential in z direction of the first layer $i = 1$ is

$$
\begin{aligned}
{}^{u,\varsigma,v}\underline{A}z,1 = {}^{u,\varsigma,v}\underline{C}_5\left(\underline{I}_{v_p}\left({}^{u,\varsigma}\underline{k}_1 r\right) + {}^{u,\varsigma,v}\underline{C}_6\underline{K}_{v_p}\left({}^{u,\varsigma}\underline{k}_1 r\right)\right) \\
{}^{\varsigma}a\cos\left(\varsigma\frac{\pi}{L_z}z\right)e^{j\left({}^{u}\omega t - v_p(\alpha+\beta)+t_u\right)}
\end{aligned}
\tag{5.21}
$$

For that the vector potential is divergence free, which is

$$
\operatorname{div}(\vec{\underline{A}}) = 0 \tag{5.22}
$$

and the radial component does not exist, we get

$$
\frac{1}{r}\frac{\partial}{\partial\alpha}{}^{u,\varsigma,v}\underline{A}_{\alpha,i} + \frac{\partial}{\partial z}{}^{u,\varsigma,v}\underline{A}_{z,i} = 0 \tag{5.23}
$$

From (5.21), the differential of vector potential $\underline{A}$ in z direction is deduced, then using integration of $\underline{A}$ in circumferential direction in (5.23), we get

$$
\begin{aligned}
{}^{u,\varsigma,v}\underline{A}_{\alpha,1} = j\frac{r\,\varsigma\pi}{v_p L_z}{}^{u,\varsigma,v}\underline{C}_5\left(\underline{I}_{v_p}\left({}^{u,\varsigma}\underline{k}_1 r\right) + {}^{u,\varsigma,v}\underline{C}_6\underline{K}_{v_p}\left({}^{u,\varsigma}\underline{k}_1 r\right)\right) \\
{}^{\varsigma}a\sin\left(\varsigma\frac{\pi}{L_z}z\right)e^{j\left({}^{u}\omega t - v_p(\alpha+\beta)+t_u\right)}
\end{aligned}
\tag{5.24}
$$

5.2.2 Vector Potential of More Layers

For every new layer, the vector potential can be calculated accordingly. Equations (5.8) and (5.9) are valid for each layer, and a setup according to (5.10) can be chosen. For each new layer a couple of ${}^{u,\varsigma,v}\underline{C}_i$ constants and one ${}^{u,\varsigma}\underline{k}_i$ constant are created. It is assumed that there are layers $i = 1,2,\ldots i_{\max}$, where $i = i_{\max}$ includes the central point where $r = 0$. The difference of each layer lies in the constants. For layers $i < i_{\max}$, we get

$$
\begin{aligned}
{}^{u,\varsigma,v}\underline{A}_{z,i} = {}^{u,\varsigma,v}\underline{C}_{3+2i}\left(\underline{I}_{v_p}\left({}^{u,\varsigma}\underline{k}_i r\right) + {}^{u,\varsigma,v}\underline{C}_{4+2i}\underline{K}_{v_p}\left({}^{u,\varsigma}\underline{k}_i r\right)\right) \\
{}^{\varsigma}a\cos\left(\varsigma\frac{\pi}{L_z}z\right)e^{j\left({}^{u}\omega t - v_p(\alpha+\beta)+t_u\right)}
\end{aligned}
\tag{5.25}
$$

$$
\begin{aligned}
{}^{u,\varsigma,v}\underline{A}_{\alpha,i} = j\frac{r\,\varsigma\pi}{v_p\,l}{}^{u,\varsigma,v}\underline{C}_{3+2i}\left(\underline{I}_{v_p}\left({}^{u,\varsigma}\underline{k}_i r\right) + {}^{u,\varsigma,v}\underline{C}_{4+2i}\underline{K}_{v_p}\left({}^{u,\varsigma}\underline{k}_i r\right)\right) \\
{}^{\varsigma}a\sin\left(\varsigma\frac{\pi}{L_z}z\right)e^{j\left({}^{u}\omega t - v_p(\alpha+\beta)+t_u\right)}
\end{aligned}
\tag{5.26}
$$

As the vector potential has to be constant for $r \to 0$ (because the magnetic field is zero at $r \to 0$), the vectors in Layer $i = i_{\max}$ are written as

$$^{u,\varsigma,v}\underline{A}_{z,i_{\max}} = {}^{u,\varsigma,v}\underline{C}_{3-2i_{\max}}\underline{I}_{v_p}\left({}^{u,\varsigma}\underline{k}_{i_{\max}}r\right)^{\varsigma}a\cos\left(\varsigma\frac{\pi}{L_z}z\right)e^{j\left({}^{u}\omega t - v_p(\alpha+\beta)+t_u\right)} \tag{5.27}$$

$$^{u,\varsigma,v}\underline{A}_{\alpha,i_{\max}} = j\frac{r}{v_p}\frac{\varsigma\pi}{L_z}{}^{u,\varsigma,v}\underline{C}_{3+2i_{\max}}\underline{I}_{v_p}\left({}^{u,\varsigma}\underline{k}_{i_{\max}}r\right)^{\varsigma}a\sin\left(\varsigma\frac{\pi}{L_z}z\right)e^{j\left({}^{u}\omega t - v_p(\alpha+\beta)+t_u\right)} \tag{5.28}$$

5.2.3 Coordinate Transformation

Layers except the stator can shield layer and airgap layer are rotating with rotor speed n. Therefore, magnetic vector potentials A of corresponding rotating layers should be transformed onto the rotor fixed coordinate that rotates synchronously. Assuming α^S and α^R representing electrical degrees respectively of the stator fixed or rotor fixed positions, the relation is written as

$$\alpha^S = \alpha^R + p\omega t = \alpha^R + 2\pi pnt \tag{5.29}$$

Considering magnetic field excitation with current harmonic u (Note that u may not apply if the magnetic field excitation does not contain harmonics), we obtain

$$\begin{aligned} {}^{u}\omega t - v_p\left(\alpha^S + \beta\right) + t_u &= {}^{u}\omega t - v_p\left(\alpha^R + 2p\pi nt + \beta\right) + t_u \\ &= {}^{u}\omega t - v_p 2p\pi nt - v_p\alpha^R - v_p\beta - t_u \\ &= \left(1 - v_p\frac{2p\pi n}{{}^{u}\omega}\right){}^{u}\omega t - v_p\alpha^R - v_p\beta + t_u \\ &= {}^{u,v}s\,{}^{u}\omega t - v_p\alpha^R - v_p\beta + t_u \\ &= {}^{u,v}s\,{}^{u}\omega t - v_p\left(\alpha^R + \beta\right) + t_u \end{aligned} \tag{5.30}$$

$$^{u,v}s = \left(1 - v_p\frac{2p\pi n}{{}^{u}\omega}\right) = 1 - v_p\left(1 - {}^{u}s\right) \tag{5.31}$$

$$^{u}s = 1 - \frac{2p\pi n}{{}^{u}\omega} = 1 - \frac{2p\pi n}{2\pi p^{u}n} = 1 - n^{u}n \tag{5.32}$$

where $^{u}\omega$ is the angular frequency of the u-th harmonic, $^{u}\omega = u\omega$, ^{u}s is slip value of the mechanical rotor, n is the speed of the rotor and ^{u}n is the synchronous speed of the u-th harmonic. Note that when SRM with DC excitation is concerned, the number of pole pairs equals the number of rotor poles, $p = N_r$. Equations (5.30), (5.31) and (5.32) mean that in the vector potential expression, instead of using $j({}^u\omega t - v_p(\alpha + \beta) + t_u)$ as exponential term of the e-function, $j({}^{u,v}s^u\omega t - v_p(\alpha + \beta) + t_u)$ is used for rotor fixed coordinate. Therefore, magnetic vector potentials of the rotating layers $i = 3,4,\ldots \leq i_{\max}$ are now alternatively written as

For $i < i_{\max}$:

$$
\begin{aligned}
{}^{u,\varsigma,v}\underline{A}_{z,i} = {}^{u,\varsigma,v}\underline{C}_{3+2i}\left(\underline{I}_{v_p}\left({}^{u,\varsigma}\underline{k}_i^{\otimes} r\right) + {}^{u,\varsigma,v}\underline{C}_{4+2i}\underline{K}_{v_p}\left({}^{u,\varsigma}\underline{k}_i^{\otimes} r\right)\right) \\
{}^{\varsigma}a \cos\left(\varsigma\frac{\pi}{L_z}z\right)e^{j\left({}^{u,v}s^u\omega t - v_p(\alpha+\beta)+t_u\right)}
\end{aligned}
\tag{5.33}
$$

$$
\begin{aligned}
{}^{u,\varsigma,v}\underline{A}_{\alpha,i} = j\frac{r\,\varsigma\pi}{v_p L_z}{}^{u,\varsigma,v}\underline{C}_{3+2i}\left(\underline{I}_{v_p}\left({}^{u,\varsigma}\underline{k}_i^{\otimes} r\right) + {}^{u,\varsigma,v}\underline{C}_{4+2i}\underline{K}_{v_p}\left({}^{u,\varsigma}\underline{k}_i^{\otimes} r\right)\right) \\
{}^{\varsigma}a \sin\left(\varsigma\frac{\pi}{L_z}z\right)e^{j\left({}^{u,v}s^u\omega t - v_p(\alpha+\beta)+t_u\right)}
\end{aligned}
\tag{5.34}
$$

For $i = i_{\max}$:

$$
{}^{u,\varsigma,v}\underline{A}_{z,i_{\max}} = {}^{u,\varsigma,v}\underline{C}_{3+2i_{\max}}\underline{I}_{v_p}\left({}^{u,\varsigma}\underline{k}_{i_{\max}}^{\otimes} r\right){}^{\varsigma}a \cos\left(\varsigma\frac{\pi}{L_z}z\right)e^{j\left({}^{u,v}s^u\omega t - v_p(\alpha+\beta)+t_u\right)} \tag{5.35}
$$

$$
{}^{u,\varsigma,v}\underline{A}_{\alpha,i_{\max}} = j\frac{r}{v_p}\frac{\varsigma\pi}{L_z}{}^{u,\varsigma,v}\underline{C}_{3+2i_{\max}}\underline{I}_{v_p}\left({}^{u,\varsigma}\underline{k}_{i_{\max}}^{\otimes} r\right){}^{\varsigma}a \sin\left(\varsigma\frac{\pi}{L_z}z\right)e^{j\left({}^{u,v}s^u\omega t - v_p(\alpha+\beta)+t_u\right)} \tag{5.36}
$$

Compared with (5.25), (5.26), (5.27) and (5.28), Eqs. (5.33), (5.34), (5.35) and (5.36) are updated with coefficient ${}^{u,v}s$ ahead of ${}^u\omega$. Meanwhile, the parameter in Bessel function ${}^{u,\varsigma}\underline{k}_i$ is now alternatively written as

$$
{}^{u,\varsigma}\underline{k}_i^{\otimes} = \sqrt{\left(\varsigma\frac{\pi}{L_z}\right)^2 + j^{u,v}s^u\omega\mu_i\gamma_i} \tag{5.37}
$$

Note that ${}^{u,\varsigma}\underline{k}_i^{\otimes}$ depends on harmonic v, and this linkage is described by a superscript defined as "⊗".

5.2.4 Calculation of Constants

Equations (5.25), (5.26), (5.27) and (5.28) and (5.33), (5.34), (5.35) and (5.36) are completely solved if the C constants are known. These constants are calculated based on boundary conditions between layers. Starting from the first layer, the magnetic field strength can be calculated from $\underline{B} = \text{rot}(\underline{A})$ respectively in α and z directions, where B is the flux density, in cylindrical coordinates.

$$^{u,\varsigma,v}\underline{B}_r = \left[\text{rot}\left(^{u,\varsigma,v}\underline{A}\right)\right]_r = \frac{1}{r}\left[\frac{\partial}{\partial\alpha}\left(^{u,\varsigma,v}\underline{A}_z\right) - \frac{\partial}{\partial z}\left(r^{u,\varsigma,v}\underline{A}_\alpha\right)\right] \quad (5.38)$$

$$^{u,\varsigma,v}\underline{B}_z = \left[\text{rot}\left(^{u,\varsigma,v}\underline{A}\right)\right]_z = \frac{1}{r}\left[\frac{\partial}{\partial r}\left(r^{u,\varsigma,v}\underline{A}_\alpha\right) - \frac{\partial}{\partial\alpha}\left(^{u,\varsigma,v}\underline{A}_r\right)\right] \quad (5.39)$$

According to (5.4), we have

$$\frac{1}{r}\frac{\partial}{\partial\alpha}{}^{u,\varsigma,v}\underline{\Theta}(\alpha,z,t) = -{}^{u,\varsigma,v}\underline{H}_{\alpha,1} = -\frac{1}{\mu_1}\frac{\partial}{\partial r}\left(^{u,\varsigma,v}\underline{A}_{z,1}\right) \quad (5.40)$$

Combining (5.40) with (5.21), we have

$$\begin{aligned}\frac{1}{r_0}\frac{\partial}{\partial\alpha}&\left[\frac{N_{ph}}{\pi v}W^v\xi_W\sqrt{2}I^\varsigma a\cos\left(\varsigma\frac{\pi}{L_z}z\right)e^{j\left(\omega t - v_p(\alpha+\beta)+{}_u\right)}\right]\\ &= \frac{1}{\mu_1}\frac{\partial}{\partial r}\left[{}^{u,\varsigma,v}\underline{C}_5\left(\underline{I}_{v_p}\left(^{u,\varsigma}\underline{k}_1 r\right) + {}^{u,\varsigma,v}\underline{C}_6\underline{K}_{v_p}\left(^{u,\varsigma}\underline{k}\ r\right)\right)\right.\\ &\left.{}^\varsigma a\cos\left(\varsigma\frac{\pi}{L_z}z\right)e^{j\left(^u\omega t - v_p(\alpha+\beta)+t_u\right)}\right]_{r=r_0}\end{aligned} \quad (5.41)$$

Further (5.41) is written as

$$-j\frac{N_{ph}}{\pi r_0}W^v\xi_W\sqrt{2}I = \frac{1}{\mu_1}{}^{u,\varsigma,v}\underline{C}_5\frac{\partial}{\partial r}\left[\underline{I}_{v_p}\left(^{u,\varsigma}\underline{k}_1 r\right) + {}^{u,\varsigma,v}\underline{C}_6\underline{K}_{v_p}\left(^{u,\varsigma}\underline{k}_1 r\right)\right]_{r=r_0} \quad (5.42)$$

Therefore, $^{u,\varsigma,v}\underline{C}_5$ is first solved and is written as

$$^{u,\varsigma,v}\underline{C}_5 = -j\mu_1\frac{N_{ph}}{\pi r_0}W^v\xi_W\sqrt{2}I\frac{1}{\frac{\partial}{\partial r}\left[\underline{I}_{v_p}\left(^{u,\varsigma}\underline{k}_1 r\right) + {}^{u,\varsigma,v}\underline{C}_6\underline{K}_{v_p}\left(^{u,\varsigma}\underline{k}_1 r\right)\right]_{r=r_0}} \quad (5.43)$$

More C constants are calculated based on boundary conditions between layers as (where $\underline{E}$ and $\underline{H}$ are electrical field strength and magnetic field intensity)

$$\begin{aligned}{}^{u,\varsigma,v}\underline{E}_{z,i}\big|_{r=r_i} = {}^{u,\varsigma,v}\underline{E}_{z,i+1}\big|_{r=r_i} \quad &\Rightarrow \quad -\frac{\partial}{\partial t}{}^{u,\varsigma,v}\underline{A}_{z,i}\big|_{r=r_i}\\ &= -\frac{\partial}{\partial t}{}^{u,\varsigma,v}\underline{A}_{z,i+1}\big|_{r=r_i}\end{aligned} \quad (5.44)$$

$$\begin{aligned}{}^{u,\varsigma,v}\underline{H}_{\alpha,i}\big|_{r=r_i} = {}^{u,\varsigma,v}\underline{H}_{\alpha,i+1}\big|_{r=r_i} \quad &\Rightarrow \quad -\frac{1}{\mu_i}\frac{\partial}{\partial r}{}^{u,\varsigma,v}\underline{A}_{z,i}\Big|_{r=r_i}\\ &= -\frac{1}{\mu_{i+1}}\frac{\partial}{\partial r}{}^{u,\varsigma,v}\underline{A}_{z,i+1}\big|_{r=r_i}\end{aligned} \quad (5.45)$$

The tangential components of the electrical field strength and magnetic field strength have to be identical on both sides of the boundary of the two layers i and $i + 1$. Starting from the innermost layer $i_{\max}$ including $r_i = 0$, ${}^{u,\varsigma,v}\underline{C}_{4+2i}$ ($i = i_{\max},\ldots,2,1$) constants of even subscripts are in sequence calculated till $i = 1$. Next in analogy, each ${}^{\varsigma,v}\underline{C}_{3+2i}$ ($i = 1,2,\ldots,i_{\max}$) constant with odd subscripts is deduced, starting from the outmost layer. According to (5.44) and (5.45) with (5.43) and the innermost layer including the point $r = 0$, as the vector potential must be constant for $r \to 0$ (See Appendix A.4), as a result,

$$i = i_{\max}: \quad {}^{u,\varsigma,v}\underline{C}_{4+2i} = 0 \tag{5.46}$$

Then the iteration to the next layer $i-1$ is obtained as

$$\begin{aligned} i &= i_{\max} - 1 : {}^{u,v,\varsigma}\underline{C}_{4+2i} \\ &= \frac{-\frac{\mu_i}{{}^{u,\varsigma}\underline{k}}\, {}_i\underline{I}_{vp}\left({}^{u,\varsigma}\underline{k}_i r_i\right)\left[\underline{I}_{vp-1}\left({}^{u,\varsigma}\underline{k}_{i+1} r_i\right)+\underline{I}_{vp+1}\left({}^{u,\varsigma}\underline{k}_{i+1} r_i\right)\right]+\frac{\mu_{i+1}}{{}^{u,\varsigma}\underline{k}}\, {}_{i+1}\underline{I}_{vp}\left({}^{u,\varsigma}\underline{k}_{i+1} r_i\right)\left[\underline{I}_{vp-1}\left({}^{u,\varsigma}\underline{k}_i r_i\right)+\underline{I}_{vp+1}\left({}^{u,\varsigma}\underline{k}_i r_i\right)\right]}{\frac{\mu_i}{{}^{u,\varsigma}\underline{k}}\, {}_i\underline{K}_{vp}\left({}^{u,\varsigma}\underline{k}_i r_i\right)\left[\underline{I}_{vp-1}\left({}^{u,\varsigma}\underline{k}_{i+1} r_i\right)+\underline{I}_{vp+1}\left({}^{u,\varsigma}\underline{k}_{i+1} r_i\right)\right]+\frac{\mu_{i+1}}{{}^{u,\varsigma}\underline{k}}\, {}_{i+1}\underline{I}_{vp}\left({}^{u,\varsigma}\underline{k}_{i+1} r_i\right)\left[\underline{K}_{vp-1}\left({}^{u,\varsigma}\underline{k}_i r_i\right)+\underline{K}_{vp+1}\left({}^{u,\varsigma}\underline{k}_i r_i\right)\right]} \end{aligned} \tag{5.47}$$

$$\begin{aligned} & i = 1, 2, \ldots, i_{\max} - 2 : \\ & {}^{u,v,\varsigma}\underline{K}_{4+2(i+1)}\left({}^{u,\varsigma}\underline{k}_{i+1} r_i\right) = \frac{{}^{u,\varsigma}\underline{k}_i}{{}^{u,\varsigma}\underline{k}_{i+1}} \frac{\mu_{i+1}}{\mu_i} \\ & \frac{\underline{I}_{v_p}\left({}^{u,\varsigma}\underline{k}_{i+1} r_i\right) + {}^{u,v,\varsigma}\underline{C}_{4+2(i+1)}\underline{K}_{v_p}\left({}^{u,\varsigma}\underline{k}_{i+1} r_i\right)}{\underline{I}_{vp-1}\left({}^{u,\varsigma}\underline{k}_{i+1} r_i\right)+\underline{I}_{vp+1}\left({}^{u,\varsigma}\underline{k}_{i+1} r_i\right)-{}^{u,v,\varsigma}\underline{C}_{4+2(i+1)}\left[\underline{K}_{vp-1}\left({}^{u,\varsigma}\underline{k}_{i+1} r_i\right)+\underline{K}_{vp+1}\left({}^{u,\varsigma}\underline{k}_{i+1} r_i\right)\right]} \end{aligned} \tag{5.48}$$

$${}^{u,v,\varsigma}\underline{C}_{4+2i} = \frac{-\underline{I}_{v_p}\left({}^{u,\varsigma}\underline{k}_i r_i\right) - \left[\underline{I}_{v_p-1}\left({}^{u,\varsigma}\underline{k}_i r_i\right) + \underline{I}_{v_p+1}\left({}^{u,\varsigma}\underline{k}_i r_i\right)\right]{}^{u,v,\varsigma}\underline{K}_{4+2(i+1)}\left({}^{u,\varsigma}\underline{k}_{i+1} r_i\right)}{\underline{K}_{v_p}\left({}^{u,\varsigma}\underline{k}_i r_i\right) + \left[\underline{K}_{v_p-1}\left({}^{u,\varsigma}\underline{k}_i r_i\right) + \underline{K}_{v_p+1}\left({}^{u,\varsigma}\underline{k}_i r_i\right)\right]{}^{u,v,\varsigma}\underline{K}_{4+2(i+1)}\left({}^{u,\varsigma}\underline{k}_{i+1} r_i\right)} \tag{5.49}$$

Next, ${}^{u,\varsigma,v}\underline{C}_{3+2i}$ constants are deduced. ${}^{u,\varsigma,v}\underline{C}_5$ from the outermost boundary layer ($i = 1$) is calculated to begin with. According to (5.43) by solving the derivative term, the following is deduced

$$\begin{aligned} & i = 1 : {}^{u,v,\varsigma}\underline{C}_{3+2i} = -j\frac{W}{\pi r_{i-1}}\,{}^{v}\xi_W \sqrt{2}\,{}^{u}I \\ & \frac{\mu_i / {}^{u,\varsigma}\underline{k}_i}{\underline{I}_{vp-1}\left({}^{u,\varsigma}\underline{k}_i r_{i-1}\right)+\underline{I}_{vp+1}\left({}^{u,\varsigma}\underline{k}_i r_{i-1}\right)-{}^{u,v,\varsigma}\underline{C}_{4+2i}\left[\underline{K}_{vp-1}\left({}^{u,\varsigma}\underline{k}_i r_{i-1}\right)+\underline{K}_{vp+1}\left({}^{u,\varsigma}\underline{k}_i r_{i-1}\right)\right]} \end{aligned} \tag{5.50}$$

Next, an iterative calculation for ${}^{u,\varsigma,v}\underline{C}_{3+2i}$ including ${}^{u,\varsigma,v}\underline{C}_{3+2i}$ and ${}^{u,\varsigma,v}\underline{C}_{3+2(i+1)}$ is carried out based on (5.25) and (5.26). The results are,

For $i = 1,2,\ldots,i_{\max} - 2$:

$$ {}^{u,v,\varsigma}\underline{C}_{3+2(i+1)} = {}^{u,v,\varsigma}\underline{C}_{3+2i}\frac{\underline{I}_{v_p}\left({}^{u,\varsigma}\underline{k}_i r_i\right) + {}^{u,v,\varsigma}\underline{C}_{4+2i}\underline{K}_{v_p}\left({}^{u,\varsigma}\underline{k}_i r_i\right)}{\underline{I}_{v_p}\left({}^{u,\varsigma}\underline{k}_{i+1} r_i\right) + {}^{u,v,\varsigma}\underline{C}_{4+2(i+1)}\underline{K}_{v_p}\left({}^{u,\varsigma}\underline{k}_{i+1} r_i\right)} \quad (5.51) $$

For $i = i_{\max} - 1$ with ${}^{u,\varsigma,v}\underline{C}_{4+2(i+1)} = 0$:

$$ {}^{u,v,\varsigma}\underline{C}_{3+2(i+1)} = {}^{u,v,\varsigma}\underline{C}_{3+2i}\frac{\underline{I}_{v_p}\left({}^{u,\varsigma}\underline{k}_i r_i\right) + {}^{u,v,\varsigma}\underline{C}_{4+2i}\underline{K}_{v_p}\left({}^{u,\varsigma}\underline{k}_i r_i\right)}{\underline{I}_{v_p}\left({}^{u,\varsigma}\underline{k}_{i+1} r_i\right)} \quad (5.52) $$

To get all ${}^{u,\varsigma,v}\underline{C}_i$ constants, the innermost layer $i = i_{\max}$ is first selected and ${}^{u,\varsigma,v}\underline{C}$ with subscript $(4 + 2i)$ is calculated. Subsequently the rest ${}^{u,\varsigma,v}\underline{C}_{4+2i}$ constants are deduced by iteration to the next outer layer. When the outer layer $i = 1$ is reached, ${}^{u,\varsigma,v}\underline{C}$ with subscript $(3 + 2i)$ are then calculated. The starting layer $i = 1$ is calculated from Ampere's Circuit Law. Similarly, the rest ${}^{u,\varsigma,v}\underline{C}_{3+2i}$ are calculated iteratively to the next inner layer, with all the known ${}^{u,\varsigma,v}\underline{C}_{4+2i}$. For each current harmonic u, the constants should be recalculated.

5.3 Magnetic Field due to DC Potential

For DC phase excitation besides sinusoidal harmonics u, there exists the potential component I_p via Fourier discretization that contributes to magnetic field as well. In this section, the magnetic vector potentials due to I_p is duduced From the method of separation of variables, the vector potential is time constant, which means the time differential

$$ {}^{u}\underline{T}'(t) = 0 \quad (5.53) $$

in (5.14) always holds true. In analogy, the differential equation is written as

$$ r^2{}^{\varsigma,v}\underline{R}''_i + r^{\varsigma,v}\underline{R}'_i - \left(\left({}^{\varsigma}\underline{k}_i\right)^2 r^2 + v_p^2\right){}^{\varsigma,v}\underline{R}_i = 0 \quad (5.54) $$

where the harmonics u in $\underline{R}_i$, and $\underline{k}_i$ as prescript are now not considered. Similarly,

$$ {}^{\varsigma}\underline{k}_i = {}^{\varsigma}\underline{k} = \sqrt{\left(\varsigma\frac{\pi}{L_z}\right)^2} = \varsigma\frac{\pi}{L_z}, \quad \varsigma = 1, 3, 5\ldots \quad (5.55) $$

which means compared with (5.19), the k_i constants for any layer makes no difference, as the alternation speed of the current potential component is 0. Therefore, the coordinate transformation is only performed on layers of rotor. The following exponent of the e-function is

$$\begin{aligned} -v_p(\alpha+\beta)+t_u &= -v_p(\alpha^R+p2\pi nt+\beta)+t_u = -v_p\alpha^R - v_p 2\pi pn - v_p\beta + t_u \\ &= -v_p\frac{2\pi pn}{\omega}\omega t - v_p\alpha^R - v_p\beta + t_u \\ &= {}^{v}s\omega t - v_p\alpha^R - v_p\beta + t_u \end{aligned} \tag{5.56}$$

$${}^{v}s = -v_p\frac{2\pi pn}{\omega} \tag{5.57}$$

As to the constants, the same deduction principle is used. Note that the superscript u for current harmonic is not necessary. The results are,

For $i = i_{\max}$:

$${}^{\varsigma,v}\underline{C}_{4+2i} = 0 \tag{5.58}$$

For $i = i_{\max} - 1$:

$$\begin{aligned} &{}^{v,\varsigma}\underline{C}_{4+2i} \\ &= \frac{-\mu_i\underline{I}_{v_p}\left({}^{\varsigma}\underline{k}r_i\right)\left[\underline{I}_{v_p-1}\left({}^{\varsigma}\underline{k}r_i\right)+\underline{I}_{v_p+1}\left({}^{\varsigma}\underline{k}r_i\right)\right]+\mu_{i+1}\underline{I}_{v_p}\left({}^{\varsigma}\underline{k}r_i\right)\left[\underline{I}_{v_p-1}\left({}^{\varsigma}\underline{k}r_i\right)+\underline{I}_{v_p+1}\left({}^{\varsigma}\underline{k}r_i\right)\right]}{\mu_i\underline{K}_{v_p}\left({}^{\varsigma}\underline{k}r_i\right)\left[\underline{I}_{v_p-1}\left({}^{\varsigma}\underline{k}r_i\right)+\underline{I}_{v_p+1}\left({}^{\varsigma}\underline{k}r_i\right)\right]+\mu_{i+1}\underline{I}_{v_p}\left({}^{\varsigma}\underline{k}r_i\right)\left[\underline{K}_{v_p-1}\left({}^{\varsigma}\underline{k}r_i\right)+\underline{K}_{v_p+1}\left({}^{\varsigma}\underline{k}r_i\right)\right]} \end{aligned} \tag{5.59}$$

For $i = 1,2,\ldots,i_{\max} - 2$:

$${}^{v,\varsigma}\underline{K}_{4+2(i+1)}\left({}^{\varsigma}\underline{k}r_i\right) = \frac{\mu_{i+1}}{\mu_i}\frac{\underline{I}_{v_p}\left({}^{\varsigma}\underline{k}r_i\right)+{}^{v,\varsigma}\underline{C}_{4+2(i+1)}\underline{K}_{v_p}\left({}^{\varsigma}\underline{k}r_i\right)}{\underline{I}_{v_p-1}\left({}^{\varsigma}\underline{k}r_i\right)+\underline{I}_{v_p+1}\left({}^{\varsigma}\underline{k}r_i\right)-{}^{v,\varsigma}\underline{C}_{4+2(i+1)}\left[\underline{K}_{v_p-1}\left({}^{\varsigma}\underline{k}r_i\right)+\underline{K}_{v_p+1}\left({}^{\varsigma}\underline{k}r_i\right)\right]} \tag{5.60}$$

$${}^{v,\varsigma}\underline{C}_{4+2i} = \frac{-\underline{I}_{v_p}\left({}^{\varsigma}\underline{k}r_i\right)-\left[\underline{I}_{v_p-1}\left({}^{\varsigma}\underline{k}r_i\right)+\underline{I}_{v_p+1}\left({}^{\varsigma}\underline{k}r_i\right)\right]{}^{v,\varsigma}\underline{K}_{4+2(i+1)}\left({}^{\varsigma}\underline{k}r_i\right)}{\underline{K}_{v_p}\left({}^{\varsigma}\underline{k}r_i\right)+\left[\underline{K}_{v_p-1}\left({}^{\varsigma}\underline{k}r_i\right)+\underline{K}_{v_p+1}\left({}^{\varsigma}\underline{k}r_i\right)\right]{}^{v,\varsigma}\underline{K}_{4+2(i+1)}\left({}^{\varsigma}\underline{k}r_i\right)} \tag{5.61}$$

For $i = 1$:

$$
{}^{\nu,\varsigma}\underline{C}_{3+2i} = -j\frac{W}{\pi r_{i-1}}{}^{\nu}\xi_W\frac{i_{DC}}{2}\frac{2\left(\mu_i/{}^{\varsigma}\underline{k}\right)}{\underline{I}_{\nu_p-1}\left({}^{\varsigma}\underline{k}r_{i-1}\right)+\underline{I}_{\nu_p+1}\left({}^{\varsigma}\underline{k}r_{i-1}\right)-{}^{\nu,\varsigma}\underline{C}_{4+2i}\left[\underline{K}_{\nu_p-1}\left({}^{\varsigma}\underline{k}r_{i-1}\right)+\underline{K}_{\nu_p+1}\left({}^{\varsigma}\underline{k}r_{i-1}\right)\right]} \tag{5.62}
$$

For $i = 1,2,\ldots,i_{\max} - 2$:

$$
{}^{\nu,\varsigma}\underline{C}_{3+2(i+1)} = {}^{\nu,\varsigma}\underline{C}_{3+2i}\frac{\underline{I}_{\nu_p}\left({}^{\varsigma}\underline{k}r_i\right)+{}^{\nu,\varsigma}\underline{C}_{4+2i}\underline{K}_{\nu_p}\left({}^{\varsigma}\underline{k}r_i\right)}{\underline{I}_{\nu_p}\left({}^{\varsigma}\underline{k}r_i\right)+{}^{\nu,\varsigma}\underline{C}_{4+2(i+1)}\underline{K}_{\nu_p}\left({}^{\varsigma}\underline{k}r_i\right)} \tag{5.63}
$$

For $i = i_{\max} - 1$ with ${}^{u,\varsigma,\nu}\underline{C}_{4+2(i+1)} = 0$:

$$
{}^{\nu,\varsigma}\underline{C}_{3+2(i+1)} = {}^{\nu,\varsigma}\underline{C}_{3+2i}\frac{\underline{I}_{\nu_p}\left({}^{\varsigma}\underline{k}r_i\right)+{}^{\nu,\varsigma}\underline{C}_{4+2i}\underline{K}_{\nu_p}\left({}^{\varsigma}\underline{k}r_i\right)}{\underline{I}_{\nu_p}\left({}^{\varsigma}\underline{k}r_i\right)} \tag{5.64}
$$

The conclusion is,
For $i = 1$,

$$
{}^{\varsigma,\nu}_{p}\underline{A}_{z,1} = {}^{\varsigma,\nu}\underline{C}_{3+2i}\left(\underline{I}_{\nu_p}\left({}^{\varsigma}\underline{k}r\right)+{}^{\varsigma,\nu}\underline{C}_{4+2i}\underline{K}_{\nu_p}\left({}^{\varsigma}\underline{k}r\right)\right){}^{\varsigma}a\cos\left(\varsigma\frac{\pi}{L_z}z\right)e^{j\left(-\frac{\nu}{p}(\alpha+\beta)+t_u\right)} \tag{5.65}
$$

$$
{}^{\varsigma,\nu}_{p}\underline{A}_{\alpha,1} = j\frac{r\,\varsigma\pi}{\nu_p L_z}{}^{\varsigma,\nu}\underline{C}_{3+2i}\left(\underline{I}_{\nu_p}\left({}^{\varsigma}\underline{k}r\right)+{}^{\varsigma,\nu}\underline{C}_{4+2i}\underline{K}_{\nu_p}\left({}^{\varsigma}\underline{k}r\right)\right){}^{\varsigma}a\sin\left(\varsigma\frac{\pi}{L_z}z\right)e^{j\left(-\nu_p(\alpha+\beta)+t_u\right)}
$$

$$
a = 0,\ \pm 1,\ \pm 2\ldots,\quad u = 1,2,3\ldots,\quad \varsigma = 1,3,5\ldots \tag{5.66}
$$

For $i = 2,3,\ldots,i_{\max} - 1$,

$$
{}^{\varsigma,\nu}_{p}\underline{A}_{z,i} = {}^{\varsigma,\nu}\underline{C}_{3+2i}\left[\underline{I}_{\nu_p}\left({}^{\varsigma}\underline{k}r\right)+{}^{\varsigma,\nu}\underline{C}_{4+2i}\underline{K}_{\nu_p}\left({}^{\varsigma}\underline{k}r\right)\right]{}^{\varsigma}a\cos\left(\varsigma\frac{\pi}{L_z}z\right)e^{j\left(-\nu_p(\alpha+\beta)+t_u\right)} \tag{5.67}
$$

$$
{}^{\varsigma,\nu}_{p}\underline{A}_{\alpha,i} = j\frac{r}{\nu_p}\frac{\varsigma\pi}{L_z}{}^{\varsigma,\nu}\underline{C}_{3+2i}\left[\underline{I}_{\nu_p}\left({}^{\varsigma}\underline{k}r\right)+{}^{\varsigma,\nu}\underline{C}_{4+2i}\underline{K}_{\nu_p}\left({}^{\varsigma}\underline{k}r\right)\right]{}^{\varsigma}a\sin\left(\varsigma\frac{\pi}{L_z}z\right)e^{j\left(-\nu_p(\alpha+\beta)+t_u\right)} \tag{5.68}
$$

For $i = i_{\max}$:

$$ {}^{\varsigma,\nu}_{p}\underline{A}_{z,i_{\max}} = {}^{\varsigma,\nu}\underline{C}_{3+2i}\underline{I}_{\nu_p}\left({}^{\varsigma}\underline{k}r\right){}^{\varsigma}a\cos\left(\varsigma\frac{\pi}{L_z}z\right)e^{j\left(-\nu_p(\alpha+\beta)+t_u\right)} \tag{5.69} $$

$$ {}^{\varsigma,\nu}_{p}\underline{A}_{\alpha,i_{\max}} = j\frac{r}{\nu_p}\frac{\varsigma\pi}{L_z}{}^{\varsigma,\nu}\underline{C}_{3+2i}\underline{I}_{\nu_p}\left({}^{\varsigma}\underline{k}r\right){}^{\varsigma}a\sin\left(\varsigma\frac{\pi}{L_z}z\right)e^{j\left(-\nu_p(\alpha+\beta)+t_u\right)} \tag{5.70} $$

Note that the prescript p is affiliated. As to calculate vector potentials in rotor fixed coordinates, instead of using $e^{j\left(-\nu_p(\alpha+\beta)+t_u\right)}$ in (5.67), (5.68), (5.69) and (5.70), the expression $e^{j\left({}^{\nu}s\omega t-\nu_p(\alpha+\beta)+t_u\right)}$ is used.

5.4 Modeling of Saliency Effect

In the concentric cylindrical layer structure model, the airgap layer is considered a smooth thin wall with constant thickness that represents magnetic reluctance. However, equivalent reluctance value is affected by saliency effect that both rotor and stator have salient poles. Due to poles that conducts magnetic flux, the reluctance values vary with rotor position. For AC excited electrical machines that have relatively more number of stator poles together with a smooth rotor, the saliency effect is not so prominent that by simply applying Carter's factor, equivalent airgap length is enlarged and seen time independent, making the airgap layer a smooth thin wall with constant thickness. However, this assumption is only valid when stator slot radial width is small compared with its height. For DC excited machines such as SRMs, both rotor and stator have prominent saliencies to generate reluctance torque, which means the equivalent reluctance value highly depends on rotor position. In addition comparatively, magnetic saturation of SRM is more obvious when the same power density is reached. Due to double saliency, the saturation is quite unevenly distributed. The distribution and intensity are regarded varying with both rotor position and phase current level, making modeling of flux characteristics that linked to the machine quite difficult. In such occasion, it is inappropriate simply to use Carter's factor to account for saliency effect.

In this concentric layer model, permeability of iron is set infinite against that of air, which means electrical steel is ideal and does not get saturated. Meanwhile, magnetic saturation from electrical steel is collectively taken as an additional part of the airgap that has an imaginary radial length that has equivalent magnetic reluctance. Therefore in this model, the total airgap includes the actual flux path between a stator and rotor pole together with such additional part, both of which are rotor position dependent.

The discretized magnetic equivalent circuit (MEC) model in Sect. 2.2 is applied to model the equivalent magnetic resistance. First, phase inductance L_{mag_ph} (θ, i), a

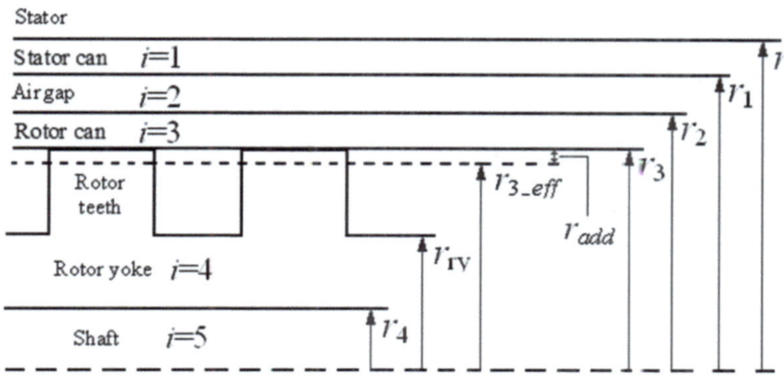

Fig. 5.5 Geometrical definition of each layer

function of rotor position θ and phase current i, is directly deduced. Then magnetic resistance of one phase $R_{mag\_ph}(\theta, i)$ is obtained by

$$R_{mag\_ph}(\theta, i) = (4N_p)^2/L_{mag\_ph}(\theta, i) \tag{5.71}$$

where N_p is the number of turns per pole. Then the equivalent airgap radial length is written as

$$l_{eff}(\theta, i) = \mu R_{mag\_ph}(\theta, i)A(\theta)/4 \tag{5.72}$$

$A(\theta)$ is defined as the effective area that flux tubes pass through airgap, which is the production of width and axial length of a pole w_{ST} and L_z respectively.

To illustrate the time dependent airgap reluctance equivalence, detailed geometry of each layer is shown in Fig. 5.5. Regardless of curvature, r_0, r_1, r_2, r_3, and r_4 are outer radius of the stator can, airgap, rotor can, rotor poles and shaft respectively, r_{ry} is radius of rotor yoke. Meanwhile, each layer is numbered. The effective airgap length excluding cans is alternatively as

$$l_{eff\_AG}(\theta, i) = l_{eff}(\theta, i) - (r_0 - r_1) - (r_2 - r_3) \tag{5.73}$$

where r_0–r_1 and r_2–r_3 are radial length of the cans respectively. It is stated here that the function $l_{eff\text{-}AG}(\theta, i)$ incorporates radial length of airgap layer, flux path deviation regulated by double saliencies as well as magnetic saturation. Therefore, the equivalent radial length of the rotor can is time varying and always higher than the nominal value r_1–r_2. The incremental value is assumed by r_{add}, a function of rotor position and phase current level, leading to numerical change of some radius of layers accordingly however. The direct solution is to reduce the rotor can radius by r_{add} to satisfy the length variation, which however will affect calculation accuracy as eddy current and loss of cans are sensitive to the geometrical change. As a simple solution, radial length of the airgap layer stays constant r_1–r_2, while that of the rotor can is time varying accordingly. The variation is modeled by setting the outer radius of the rotor can constant r_2 while the inner radius $r_{3\_eff} = r_3 - r_{add}$. In other words, the saliency effect is taken collectively as one part of the inner can expressed by r_{add} without change of the airgap layer. Such transfer is feasible, as the can material is non-magnetic. However, radial length of the rotor can is increased, leading to overestimation of the can loss from the rotor can. To overcome this problem, a coefficient is headed as a means of correction for rotor can loss calculation. Based on that the loss is proportional to radial length of the can layer, the coefficient a is expressed as a ratio below

$$a = (r_2 - r_3)/\left(r_2 - r_{3\_eff}\right) \tag{5.74}$$

5.5 Feature Deduction

5.5.1 *Flux Density of Stator*

For machines free from rotor magnetization, flux density excitation originates from the armature coils of stator only. For permanent magnet machines, additional source comes from PMs. In the following, flux density of each source is deduced, including poles and yoke.

5.5.1.1 Stator Yoke

For stator armature coils, based on (5.21) and (5.24), magnetic flux density at the stator surface $i = 1$ is written as (Note that the superscript "S" stands for stator current.)

$$\begin{aligned}
{}^{u,\varsigma,v}\underline{B}^{S}_{r,1}(r,\alpha)\big|_{r=r_0} &= \left[\tfrac{1}{r}\left(\tfrac{\partial}{\partial\alpha}\left({}^{u,\varsigma,v}\underline{A}^{S}_{z,1}\right) - \tfrac{\partial}{\partial z}\left(r\,{}^{u,\varsigma,v}\underline{A}^{S}_{\alpha,1}\right)\right)\right]\Big|_{r=r_0} \\
&= -j\frac{r_0}{v_p}\left(\left(\frac{v_p}{r_0}\right)^2 + \left(\frac{\varsigma\pi}{L_z}\right)^2\right){}^{u,\varsigma,v}\underline{C}_5\left(\underline{I}_{v_p}\left({}^{u,\varsigma}\underline{k}_1 r_0\right) + {}^{u,\varsigma,v}\underline{C}_6\underline{K}_{v_p}\left({}^{u,\varsigma}\underline{k}_1 r_0\right)\right) \\
&\qquad {}^{\varsigma}a\sin\left(\varsigma\frac{\pi}{L_z}z\right)e^{j\left({}^{u}\omega t - v_p(\alpha+\beta)+t_u\right)}
\end{aligned} \tag{5.75}$$

For PMs as magnetization source (if applicable) [16], based on Sect. 4.8, in analogy we obtain (Note that superscript "m" stands for magnets, κ and ξ are axial and circumferential harmonics for magnets.)

$$ {}^{\kappa,\xi}\underline{B}^{m}_{r,1}(r,\alpha)\Big|_{r=r_0} = \left[\frac{1}{r}\left(\frac{\partial}{\partial\alpha}\left({}^{\kappa,\xi}\underline{A}^{m}_{z,1}\right) - \frac{\partial}{\partial z}\left(r^{u,\varsigma,\nu}\underline{A}^{m}_{\alpha,1}\right)\right)\right]\Big|_{r=r_0} \tag{5.76} $$

For machines free from rotor magnetization,

$$ {}^{\kappa,\xi}\underline{B}^{m}_{r,1}(r,\alpha)\Big|_{r=r_0} = 0 \tag{5.77} $$

The total flux density on stator surface due to magnetic field of permanent magnets and magnetic field of the stator currents can be obtained as sum of (5.75) and (5.76) or (5.77) as follows

$$ \begin{aligned} &{}^{u,\varsigma,\nu,\kappa,\xi}\underline{B}_{r,1}(r,\alpha)\Big|_{r=r_0} \\ &= \left({}^{u,\varsigma,\nu}\underline{B}^{S}_{r,1}(r,\alpha) + {}^{\kappa,\xi}\underline{B}^{m}_{r,1}(r,\alpha)\right)\Big|_{r=r_0} \end{aligned} \tag{5.78} $$

By integration over the axis $-L_z/2 \le z \le L_z/2$ and circumference $0 \le \alpha \le 2\pi$, the flux values are obtained, which is then divided by cross section of stator yoke $A_{S,yoke}$. Also considering insulation between lamination sheets by the filling factor k_{Fe}, the airgap flux density is written as

$$ {}^{u,\varsigma,\nu,\kappa,\xi}\underline{B}_{S,yoke} = \frac{1}{2k_{Fe}A_{S,yoke}} \int_{-L_z/2}^{L_z/2}\int_{0}^{2\pi} {}^{u,\varsigma,\nu,\kappa,\xi}\underline{B}_{r,1}(r,\alpha)\Big|_{r=r_0} d\alpha dz \tag{5.79} $$

5.5.1.2 Stator Pole

Regardless of leakage, all airgap magnetic flux is assumed ideally to pass through corresponding stator pole and flux density of stator poles is obtained by integrating over the surface under one stator slot pitch τ_N that is then divided by the cross section of stator pole $A_{S,pole}$ and multiplied by the filling factor k_{Fe}. Note that π/τ_p converts the corresponding length into electrical degrees. The equation is written as

$$ {}^{u,\varsigma,\nu,\kappa,\xi}\underline{B}_{S,pole} = \frac{1}{k_{Fe}A_{S,yoke}} \int_{-\frac{L_z}{2}}^{\frac{L_z}{2}}\int_{-\frac{\tau_N}{2}\frac{\pi}{\tau_p}}^{\frac{\tau_N}{2}\frac{\pi}{\tau_p}} {}^{u,\varsigma,\nu,\kappa,\xi}\underline{B}^{S}_{r,1}(r,\alpha)\Big|_{r=r_0} d\alpha dz \tag{5.80} $$

5.5.2 Flux Linkage

According to [15, 17], to calculate flux that is linked to stator coils, we start from a single basic excitation unit. Based on MMF models in Chap. 4, for Model 1 of tooth concentrated coils, the unit is γ-th layer of ρ-th coil of k-th phase while for Model 2, the unit is ρ-th coil in one phase and k-th phase. For AC machines, it is κ-th coil of ρ-th coil group of k-th phase. For similarity, we do not consider magnets (The same principle applies in the following.) and therefore the superscripts "S" and "m" respectively for stator or magnet excitation is neglected. Here Model 1 with wire layout is investigated as example. For one unit, we consider the integration of flux density over the spatial space it takes. Specifically in circumferential direction the range is $[\alpha_1, \alpha_2]$ while in axial direction it is ${}^{\varsigma}a{\cdot}\cos(\varsigma(\pi/L_z)\mathrm{z})$. Therefore, flux linkage Ψ due to γ-th layer of ρ-th coil of k-th phase is written as

$$ {}^{u,\varsigma,v}\underline{\Psi}_{\gamma,\rho,k} = -\frac{N_c}{S}\int_{-\frac{L_z}{2}}^{\frac{L_z}{2}}\int_{\alpha_2}^{\alpha_1} {}^{\varsigma}a\cos\left(\varsigma\frac{\pi}{L_z}z\right){}^{u,\varsigma,v}\underline{B}_{r,1}(r,\alpha)\bigg|_{r=r_0} r_0 \mathrm{d}\alpha \mathrm{d}z \tag{5.81} $$

N_c is the number of turns per coil. Note that there is a negative sign, as the direction of the r-coordinate is opposite to the unit vector of stator surface (See Figs. 5.1, 5.2 and 5.3). The circumferential position α_1 and α_2 characterizes the start and end of a coil. Equation (5.81) combining (5.76) can be further written as

$$ {}^{u,\varsigma,v}\underline{\Psi}_{\gamma,\rho,k} = \frac{N_c}{S}\int_{-\frac{L_z}{2}}^{\frac{L_z}{2}}\int_{\alpha_2}^{\alpha_1} \frac{r_0^2}{v_p}\left(\left(\frac{v_p}{r_0}\right)^2+\left(\frac{\varsigma\pi}{L_z}\right)^2\right)^2 {}^{u,\varsigma,v}\underline{C}_5\left(\underline{I}_{v_p}\left({}^{u,\varsigma}\underline{k}_1 r_0\right)+{}^{u,\varsigma,v}\underline{C}_6\underline{K}_{v_p}\left({}^{u,\varsigma}\underline{k}_1 r_0\right)\right) \left({}^{\varsigma}a\cos\left(\varsigma\frac{\pi}{L_z}z\right)\right)^2 e^{j\left({}^{u}\omega t - v_p(\alpha+\beta)+t_u\right)}\mathrm{d}\alpha\mathrm{d}z \tag{5.82} $$

where is the number of turns per coil and S is the number of layers within a coil, N_c/S means number of turns per layer. Then (5.82) is further written as

$$ {}^{u,\varsigma,v}\underline{\Psi}_{\gamma,\rho,k} = \frac{N_c\left({}^{\varsigma}ar_0\right)^2}{Sv_p}\left(\left(\frac{v_p}{r_0}\right)^2+\left(\frac{\varsigma\pi}{L_z}\right)^2\right)^2 {}^{u,\varsigma,v}\underline{C}_5\left(\underline{I}_{v_p}\left({}^{u,\varsigma}\underline{k}_1 r_0\right)+{}^{u,\varsigma,v}\underline{C}_6\underline{K}_{v_p}\left({}^{u,\varsigma}\underline{k}_1 r_0\right)\right) e^{j\left({}^{u}\omega t+t_u\right)}\int_{-\frac{L_z}{2}}^{\frac{L_z}{2}}\left(\cos\left(\varsigma\frac{\pi}{L_z}z\right)\right)^2\mathrm{d}z\int_{\alpha_2}^{\alpha_1} e^{-jv_p(\alpha+\beta)}\mathrm{d}\alpha \tag{5.83} $$

The first integral in (5.83) leads to

$$\int_{-\frac{L_z}{2}}^{\frac{L_z}{2}} \left(\cos\left(\varsigma \frac{\pi}{L_z} z \right) \right)^2 \mathrm{d}z = \left[\frac{1}{2} z + \frac{1}{4\varsigma \frac{\pi}{L_z}} \sin\left(2\varsigma \frac{\pi}{L_z} z \right) \right]_{z=-\frac{L_z}{2}}^{z=\frac{L_z}{2}}$$
$$= \frac{L_z}{2} + \frac{L_z}{4\varsigma\pi}[\sin(\varsigma\pi) - \sin(-\varsigma\pi)] = \frac{L_z}{2} \tag{5.84}$$

The second integral in (5.83) leads to

$$\int_{\alpha_2}^{\alpha_1} e^{-jv_p(\alpha+\beta)} \mathrm{d}\alpha = \int_{-\frac{w_{co}}{2}\frac{\pi}{\tau_p}}^{\frac{w_{co}}{2}\frac{\pi}{\tau_p}} e^{-jv_p(\alpha+\beta)} \mathrm{d}(\alpha+\beta) = \frac{2}{v_p} \sin\left(v_p \frac{w_{co}\pi}{\tau_p\, 2} \right) \tag{5.85}$$

where w_{co} is the width between two sides of a coil defined in Fig. 4.5 in Chap. 4 Then (5.83) leads to

$$^{u,\varsigma,v}\underline{\Psi}_{\gamma,\rho,k} = \frac{N_c({}^{\varsigma}a r_0)^2 L_z}{S v_p^2} \left(\left(\frac{v_p}{r_0}\right)^2 + \left(\frac{\varsigma\pi}{L_z}\right)^2 \right)^2$$
$$^{u,\varsigma,v}\underline{C}_5 \left(\underline{I}_{v_p}\left({}^{u,\varsigma}\underline{k}_1 r_0\right) + {}^{u,\varsigma,v}\underline{C}_6 \underline{K}_{v_p}\left({}^{u,\varsigma}\underline{k}_1 r_0\right) \right) \sin\left(v_p \frac{w_{co}\pi}{\tau_p\, 2} \right) e^{j\left({}^u\omega t + t_u\right)} \tag{5.86}$$

Now the flux linkage is calculated by sum of all γ layers then all ρ coils and finally all k phases, which is

$$^{u,\varsigma,v}\underline{\Psi} = \sum_k \sum_\rho \sum_\gamma \frac{N_c({}^{\varsigma}a r_0)^2 L_z}{S v_p^2} \left(\left(\frac{v_p}{r_0}\right)^2 + \left(\frac{\varsigma\pi}{L_z}\right)^2 \right)^2$$
$$^{u,\varsigma,v}\underline{C}_5 \left(\underline{I}_{v_p}\left({}^{u,\varsigma}\underline{k}_1 r_0\right) + {}^{u,\varsigma,v}\underline{C}_6 \underline{K}_{v_p}\left({}^{u,\varsigma}\underline{k}_1 r_0\right) \right) \sin\left(v_p \frac{w_{co}\pi}{\tau_p\, 2} \right) e^{j\left({}^u\omega t + t_u\right)} \tag{5.87}$$

According to the summation deduction in Chap. 4, we get S, ${}^v\xi_Z$, $2p$ and the sinusoidal term in (4.84). Such deduction is again applied in (5.87). Note that for SRMs the number of pole pairs p equals to the number of rotor poles N_r. Then (5.87) leads to

$$^{u,\varsigma,v}\underline{\Psi} = \frac{W({}^{\varsigma}a r_0)^{2\,v}\xi_Z L_z}{S v_p^2} \left(\left(\frac{v_p}{r_0}\right)^2 + \left(\frac{\varsigma\pi}{L_z}\right)^2 \right)^2 \frac{\sin\left(\left(\frac{v}{p} - 2u \right) \frac{p\pi}{4} \right)}{\sin\left(\left(\frac{v}{p} - 2u \right) \frac{p\pi}{4N_{ph}} \right)}$$
$$^{u,\varsigma,v}\underline{C}_5 \left(\underline{I}_{v_p}\left({}^{u,\varsigma}\underline{k}_1 r_0\right) + {}^{u,\varsigma,v}\underline{C}_6 \underline{K}_{v_p}\left({}^{u,\varsigma}\underline{k}_1 r_0\right) \right) \sin\left(v_p \frac{w_{co}\pi}{\tau_p\, 2} \right) e^{j\left({}^u\omega t + t_u\right)} \tag{5.88}$$

5.5.3 Iron Loss

Iron loss calculation in magnetic materials is traditionally uses sinusoidal flux density of varying frequency and magnitude. Usually under alternating flux conditions the iron loss density p_{Fe} in W/kg is separated into hysteresis loss component p_{hys} and eddy current component p_{edd}. The iron loss power density is given by [18]

$$p_{Fe} = p_{hys} + p_{edd} = \left(c_{hys}\frac{f}{50} + c_{edd}\left(\frac{f}{50}\right)^{x_1} \right)\widehat{B}^{x_2} \tag{5.89}$$

where f is the excitation frequency, $\widehat{B}$ is the amplitude value of the sinusoidal flux density, c_{hys} and c_{edd} are hysteresis and eddy current constants, x_1 and x_2 are coefficients depending on core material. The total iron loss P_{Fe} is written as

$$p_{Fe} = \left(p_{hys} + p_{edd}\right)\Big|_{f,B_{pole}} G_{pole} + \left(p_{hys} + p_{edd}\right)\Big|_{f,B_{yoke}} G_{yoke} \tag{5.90}$$

where B_{pole} and B_{yoke} are flux density of poles and yoke, G_{pole} and G_{yoke} are weight of poles and yoke respectively. The iron loss expression in (5.90) is valid only for sinusoidal flux density excitation. In most cases the variation in flux density in the core is far from sinusoidal. In this situation, while the hysteresis loss is still easy to evaluate as it depends only on the peak value of the flux density assuming that there are no minor hysteresis loops, eddy current losses evaluated using only the fundamental component of flux density may be much lower than the measured values. For eddy current it is convenient to represent the average loss density as a function of rate of time change of the vector flux density dB/dt [19, 20], where T is the time period.

$$p_{edd} = \frac{2c_{edd}}{T}\int_0^T \left(\frac{\mathrm{d}B}{\mathrm{d}t}\right)^2 \mathrm{d}t \tag{5.91}$$

In addition, anomalous loss as another part is also taken. Based on [20], an improved analytical model for core losses that accounts for stator armature reaction is presented in [21]. The losses in poles and yoke can be determined from the followings as

$$P_{Fe\_pole} = \left\{ c_{hys}\widehat{B}^2_{pole}f + \left[\frac{1}{T}\int_0^T c_{edd}\left(\frac{\mathrm{d}B_{pole}}{\mathrm{d}t}\right)^2 \mathrm{d}t\right] + \left[\frac{1}{T}\int_0^T c_{exc}\left(\frac{\mathrm{d}B_{pole}}{\mathrm{d}t}\right)^{3/2} \mathrm{d}t\right] \right\} G_{pole} \tag{5.92}$$

$$P_{Fe\_yoke} = \left\{ c_{hys}\widehat{B}^2_{yoke}f + \left[\frac{1}{T}\int_0^T c_{edd}\left(\frac{dB_{yoke}}{dt}\right)^2 dt\right] + \left[\frac{1}{T}\int_0^T c_{exc}\left(\frac{dB_{yoke}}{dt}\right)^{3/2} dt\right]\right\} G_{yoke} \tag{5.93}$$

where c_{exc} is the excess loss constant. Flux density of each component is calculated according to Sect. 5.5.1 or by the obtained magnetic vector potentials. Note that $^{u,v}s$ is used concerning the flux density expressions of rotating layers when calculating rotor loss.

5.5.4 Can Loss

Loss of each layer including can shields is calculated by Poynting's theorem of energy conversion in electromagnetic field. The energy vector is written as

$$\underline{\vec{S}} = \underline{\vec{E}} \times \underline{\vec{H}} \tag{5.94}$$

where S is Poynting's vector representing directional energy flux density of the electromagnetic field, E and H are respectively the electrical field density and magnetic field intensity, the vector form is arrowed and complex form is underlined. For time periodic sinusoidal excited fields, the average power flow is more useful. The Poynting's vector can be further written into a couple of parts as

$$\begin{aligned}\underline{\vec{S}} &= \underline{\vec{E}} \times \underline{\vec{H}} = \frac{1}{2}\left(\underline{\vec{E}}e^{j\omega t} + \underline{\vec{E}}^{*} e^{-j\omega t}\right) \times \frac{1}{2}\left(\underline{\vec{H}}e^{j\omega t} + \underline{\vec{H}}^{*} e^{-j\omega t}\right) \\ &= \frac{1}{2}\mathrm{Re}\left\{\underline{\vec{E}} \times \underline{\vec{H}}^{*}\right\} + \frac{1}{2}\mathrm{Re}\left\{\underline{\vec{E}} \times \underline{\vec{H}}e^{2j\omega t}\right\} = \underline{\vec{S}}_{ave} + \underline{\vec{S}}_{ins} = \underline{\vec{S}}_{ave} + \underline{\vec{S}}(t)\end{aligned} \tag{5.95}$$

where the superscript "*" is conjugate operator, the subscripts "ins" and "ave" mean instantaneous and averaged value respectively. As the average value of the second part is zero, the time average energy flow is written as

$$\underline{\vec{S}}_{ave} = \frac{1}{2}\mathrm{Re}\left\{\underline{\vec{E}} \times \underline{\vec{H}}^{*}\right\} \tag{5.96}$$

The energy flows by each current harmonic are summed up and loss is obtained by integrating the outer surface Ω of each layer, which is written as

$$P_{ave} = -\frac{1}{2}\iint_{\Omega} \mathrm{Re}\left\{\underline{\vec{E}} \times \underline{\vec{H}}^{*}\right\} \vec{n}\, d\Omega \tag{5.97}$$

where minus sign considers the sink of power as a positive value, as the power source is positive according to Gauss theorem, n is the unit vector vertical to Ω. Solving the surface integral in stator fixed coordinates via Gauss theorem, the time average active power delivered to each layer is obtained. Based on the regarded coordinate system of each layer, ohmic loss value is obtained. According to Appendix A.5, the loss of each layer is written as.

For $i = 1, 2, 3, \ldots, i_{\max}-1$:

$$\begin{aligned} P_{i\_ave} = & -\frac{\pi L_z}{4\mu_i}\sum_u\sum_\varsigma\sum_v {}^\varsigma a^2 \left|{}^{u,\varsigma,v}\underline{C}_{3+2i}\right|^2 \frac{{}^u\omega}{v_p^2} \\ & \mathrm{Im}\left\{ {}^{u,\varsigma}\underline{k}_i^* r_{i-1}^3 \left[\left(\frac{\varsigma\pi}{L_z}\right)^2 + \left(\frac{v_p}{r_{i-1}}\right)^2\right]\left[\underline{I}_{v_p}\left({}^{u,\varsigma}\underline{k}_i r_{i-1}\right) + {}^{u,\varsigma,v}\underline{C}_{4+2i}\underline{K}_{v_p}\left({}^{u,\varsigma}\underline{k}_i r_{i-1}\right)\right] \right. \\ & \times\left[\underline{I}_{v_p-1}\left({}^{u,\varsigma}\underline{k}_i^* r_{i-1}\right) + \underline{I}_{v_p+1}\left({}^{u,\varsigma}\underline{k}_i^* r_{i-1}\right) - {}^{u,\varsigma,v}\underline{C}_{4+2i}^*\left(\underline{K}_{v_p-1}\left({}^{u,\varsigma}\underline{k}_i^* r_{i-1}\right) + \underline{K}_{v_p+1}\left({}^{u,\varsigma}\underline{k}_i^* r_{i-1}\right)\right)\right] \\ & - {}^{u,\varsigma}\underline{k}_i^* r_i^3 \left[\left(\frac{\varsigma\pi}{L_z}\right)^2 + \left(\frac{v_p}{r_i}\right)^2\right]\left[\underline{I}_{v_p}\left({}^{u,\varsigma}\underline{k}_i r_i\right) + {}^{u,\varsigma,v}\underline{C}_{4+2i}\underline{K}_{v_p}\left({}^{u,\varsigma}\underline{k}_i r_i\right)\right] \\ & \left. \times\left[\underline{I}_{v_p-1}\left({}^{u,\varsigma}\underline{k}_i^* r_i\right) + \underline{I}_{v_p+1}\left({}^{u,\varsigma}\underline{k}_i^* r_i\right) - {}^{u,\varsigma,v}\underline{C}_{4+2i}^*\left(\underline{K}_{v_p-1}\left({}^{u,\varsigma}\underline{k}_i^* r_i\right) + \underline{K}_{v_p+1}\left({}^{u,\varsigma}\underline{k}_i^* r_i\right)\right)\right]\right\} \end{aligned} \tag{5.98}$$

For $i = i_{\max}$:

$$\begin{aligned} P_{i\_ave} = & -\frac{\pi L_z}{4\mu_i}\sum_u\sum_\varsigma\sum_v {}^\varsigma a^2 \left|{}^{u,\varsigma,v}\underline{C}_{3+2i}\right|^2 \frac{{}^u\omega}{v_p^2} \\ & \mathrm{Im}\left\{ {}^{u,\varsigma}\underline{k}_i^* r_{i-1}^3 \left(\left(\frac{\varsigma\pi}{L_z}\right)^2 + \left(\frac{v_p}{r_{i-1}}\right)^2\right)\underline{I}_{v_p}\left({}^{u,\varsigma}\underline{k}_i r_{i-1}\right)\left(\underline{I}_{v_p-1}\left({}^{u,\varsigma}\underline{k}_i^* r_{i-1}\right) + \underline{I}_{v_p+1}\left({}^{u,\varsigma}\underline{k}_i^* r_{i-1}\right)\right)\right\} \end{aligned} \tag{5.99}$$

When the rotor can loss is calculated, the coefficient a in (5.74) as a means of correction should be headed in (5.98). Also to get the power loss in the rotor and rotor can, it is necessary to transform the above equations to the rotating reference frame. Equations (5.97) and (5.98) are transformed according to the following principles,

(a) ${}^u\omega$ is replaced by ${}^{u,v}s\ {}^u\omega$.
(b) ${}^{u,\varsigma}\underline{k}_i$and ${}^{u,\varsigma}\underline{k}_i^*$ are replaced by ${}^{u,\varsigma}\underline{k}_i^\otimes$ and ${}^{u,\varsigma}\underline{k}_i^{\otimes *}$ respectively.

5.5.5 *Torque*

Torque is obtained by sum of mechanical power of each layer, or simply the negative torque from the stator, according to Newton's Law ($T_{rotor} = -T_{stator}$). To get the torque on the stator, calculation of airgap flux density is necessary. The stator torque is obtained by applying Maxwell's stress tensor in cylindrical coordinates on stator surface. The mechanical force density of u-th harmonic in circumferential direction is written as

$$^{u,\varsigma,v}f_{\alpha} = -\frac{1}{2}\mathrm{Re}\left\{\frac{1}{\mu_1}{}^{u,\varsigma,v}\underline{B}_{\alpha,1}{}^{u,\varsigma,v}\underline{B}^{*}_{r,1}\right\}\Big|_{r=r_0} \tag{5.100}$$

The torque is calculated by sum of current harmonics as

$$\begin{aligned} -T_{stator} &= -\sum_{u}\sum_{\varsigma}\sum_{v}\int_{-\pi}^{\pi}\int_{0}^{r_0}\int_{-\frac{L_z}{2}}^{\frac{L_z}{2}}{}^{u,\varsigma,v}f_{\alpha}\mathrm{d}r\mathrm{d}zr\mathrm{d}\alpha \\ &= \frac{r_0^2}{2\mu_1}\sum_{u}\sum_{\varsigma}\sum_{v}\int_{-\pi}^{\pi}\int_{-\frac{L_z}{2}}^{\frac{L_z}{2}}\mathrm{Re}\left\{{}^{u,\varsigma,v}\underline{B}_{\alpha,1}{}^{u,\varsigma,v}\underline{B}^{*}_{r_0,1}\right\}\mathrm{d}z\mathrm{d}\alpha \end{aligned} \tag{5.101}$$

Equation (5.101) is further deduced with the following equations as

$$\begin{aligned} {}^{u,\varsigma,v}\underline{B}_{\alpha,1} &= \left(\mathrm{rot}\left({}^{u,\varsigma,v}\underline{A}\right)\right)_{\alpha,1} = -\frac{\partial}{\partial r}\left({}^{u,\varsigma,v}\underline{A}_{z,1}\right) \\ &= -{}^{u,\varsigma,v}\underline{C}_5\left[\begin{array}{l}\underline{I}_{v_p-1}\left({}^{u,\varsigma}\underline{k}_1 r\right) + \underline{I}_{v_p+1}\left({}^{u,\varsigma}\underline{k}_1 r\right) \\ -{}^{u,\varsigma,v}\underline{C}_6\underline{K}_{v_p-1}\left({}^{u,\varsigma}\underline{k}_1 r_0\right) + \underline{K}_{v_p+1}\left({}^{u,\varsigma}\underline{k}_1 r_0\right)\end{array}\right] \\ &\quad \frac{1}{2}{}^{u,\varsigma}\underline{k}_1{}^{\varsigma}a\cos\left(\varsigma\frac{\pi}{L_z}z\right)e^{j\left({}^{u}\omega t - v_p(\alpha+\beta)+t_u\right)} \end{aligned} \tag{5.102}$$

$$^{u,\varsigma,v}\underline{B}^{*}_{r,1} = \left[\mathrm{rot}\left({}^{u,\varsigma,v}\underline{A}\right)\right]^{*}_{r,1} = \frac{1}{r}\left[\frac{\partial}{\partial\alpha}\left({}^{u,\varsigma,v}\underline{A}_{z,1}\right) - \frac{\partial}{\partial z}\left(r^{u,\varsigma,v}\underline{A}_{\alpha,1}\right)\right]^{*} \tag{5.103}$$

where

$$\begin{aligned} \frac{1}{r}\frac{\partial}{\partial\alpha}\left({}^{u,\varsigma,v}\underline{A}_{z,1}\right) = -j\frac{v_p}{r}{}^{u,\varsigma,v}\underline{C}_5\left[\underline{I}_{v_p}\left({}^{u,\varsigma}\underline{k}_1 r\right) + {}^{u,\varsigma,v}\underline{C}_6\underline{K}_{v_p}\left({}^{u,\varsigma}\underline{k}_1 r\right)\right] \\ {}^{\varsigma}a\cos\left(\varsigma\frac{\pi}{L_z}z\right)e^{j\left({}^{u}\omega t - v_p(\alpha+\beta)+t_u\right)} \end{aligned} \tag{5.104}$$

$$\begin{aligned} \frac{\partial}{\partial z}\left({}^{u,\varsigma,v}\underline{A}_{\alpha,1}\right) = j\frac{r}{v_p}\left(\frac{\varsigma\pi}{L_z}\right)^2{}^{u,\varsigma,v}\underline{C}_5\left[\underline{I}_{v_p}\left({}^{u,\varsigma}\underline{k}_1 r\right) + {}^{u,\varsigma,v}\underline{C}_6\underline{K}_{v_p}\left({}^{u,\varsigma}\underline{k}_1 r\right)\right] \\ {}^{\varsigma}a\cos\left(\varsigma\frac{\pi}{L_z}z\right)e^{j\left({}^{u}\omega t - v_p(\alpha+\beta)+t_u\right)} \end{aligned} \tag{5.105}$$

Based on (5.103), (5.104) and (5.105), (5.102) is further written as

$$
{}^{u,\varsigma,v}\underline{B}_{r,1}^{*} = j\frac{r}{v_p}\left(\left(\frac{\varsigma\pi}{L_z}\right)^2 + \left(\frac{v_p}{r}\right)^2\right){}^{u,\varsigma,v}\underline{C}_5^{*}\left[\underline{I}_{v_p}\left({}^{u,\varsigma}\underline{k}_1^{*}r\right) + {}^{u,\varsigma,v}\underline{C}_6^{*}\underline{K}_{v_p}\left({}^{u,\varsigma}\underline{k}_1^{*}r\right)\right]
{}^{\varsigma}a\cos\left(\varsigma\frac{\pi}{L_z}z\right)e^{-j\left({}^{u}\omega t - v_p(\alpha+\beta)+t_u\right)} \tag{5.106}
$$

Equation (5.101) and (5.106) lead to

$$
\mathrm{Re}\left\{{}^{u,\varsigma,v}\underline{B}_{\alpha,1}\,{}^{u,\varsigma,v}\underline{B}_{r,1}^{*}\right\}\Big|_{r=r_0} = \frac{1 r_0}{2v_p}\left(\left(\frac{\varsigma\pi}{L_z}\right)^2 + \left(\frac{v_p}{r_0}\right)^2\right){}^{\varsigma}a^2\cos^2\left(\varsigma\frac{\pi}{L_z}z\right)\left|{}^{u,\varsigma,v}\underline{C}_5\right|^2
\mathrm{Re}\left\{\begin{array}{l} -j^{u,\varsigma}\underline{k}_1\left[\begin{array}{l}\underline{I}_{v_p-1}\left({}^{u,\varsigma}\underline{k}_1 r_0\right) + \underline{I}_{v_p+1}\left({}^{u,\varsigma}\underline{k}_1 r_0\right) - \\ {}^{u,\varsigma,v}\underline{C}_6\left[\underline{K}_{v_p-1}\left({}^{u,\varsigma}\underline{k}_1 r_0\right) + \underline{K}_{v_p+1}\left({}^{u,\varsigma}\underline{k}_1 r_0\right)\right]\end{array}\right] \\ \left[\underline{I}_{v_p}\left({}^{u,\varsigma}\underline{k}_1^{*} r_0\right) + {}^{u,\varsigma,v}\underline{C}_6^{*}\underline{K}_{v_p}\left({}^{u,\varsigma}\underline{k}_1^{*} r_0\right)\right]\end{array}\right\} \tag{5.107}
$$

Using the integral,

$$
\int_{-\pi}^{\pi} \mathrm{d}\alpha = 2\pi \tag{5.108}
$$

$$
\int_{-\frac{L_z}{2}}^{\frac{L_z}{2}} \cos^2\left(\frac{\varsigma\pi}{L_z}z\right)\mathrm{d}z = \frac{L_z}{2} \tag{5.109}
$$

Torque by phase current harmonics is written as

$$
-T_{stator} = \frac{\pi r_0^3}{4\mu_1}L_z\sum_u\sum_\varsigma\sum_v\frac{1}{v_p}\left[\left(\frac{\varsigma\pi}{L_z}\right)^2 + \left(\frac{v_p}{r_0}\right)^2\right]{}^{\varsigma}a^2\left|{}^{u,\varsigma,v}\underline{C}_5\right|^2
\mathrm{Im}\left\{\begin{array}{l} {}^{u,\varsigma}\underline{k}_1\left[\begin{array}{l}\underline{I}_{v_p-1}\left({}^{u,\varsigma}\underline{k}_1 r_0\right) + \underline{I}_{v_p+1}\left({}^{u,\varsigma}\underline{k}_1 r_0\right) \\ -{}^{u,\varsigma,v}\underline{C}_6\left[\underline{K}_{v_p-1}\left({}^{u,\varsigma}\underline{k}_1 r_0\right) + \underline{K}_{v_p+1}\left({}^{u,\varsigma}\underline{k}_1 r_0\right)\right]\end{array}\right] \\ \left[\underline{I}_{v_p}\left({}^{u,\varsigma}\underline{k}_1^{*} r_0\right) + {}^{u,\varsigma,v}\underline{C}_6^{*}\underline{K}_{v_p}\left({}^{u,\varsigma}\underline{k}_1^{*} r_0\right)\right]\end{array}\right\} \tag{5.110}
$$

When cans are inserted in the air gap, the torque is the sum of all stator fixed layers, including the stator bore and the outer can cylinder, which is

$$
T = -(T_{stator} + T_{rotorcan}) \tag{5.111}
$$

Similarly, mechanical torque from the stator can shield is written as

$$-T_{statorcan} = \frac{\pi r_1^3}{4\mu_2} L_z \sum_u \sum_\varsigma \sum_v \frac{1}{v_p}\left[\left(\frac{\varsigma\pi}{L_z}\right)^2 + \left(\frac{v_p}{r_1}\right)^2\right] {}^{\varsigma}a^2 \left|{}^{u,\varsigma,v}\underline{C}_7\right|^2$$
$$\mathrm{Im}\left\{ {}^{u,\varsigma}\underline{k}_2 \left[\begin{array}{l} \underline{I}_{v_p-1}\left({}^{u,\varsigma}\underline{k}_2 r_1\right) + \underline{I}_{v_p+1}\left({}^{u,\varsigma}\underline{k}_2 r_1\right) \\ -{}^{u,\varsigma,v}\underline{C}_8\left[\underline{K}_{v_p-1}\left({}^{u,\varsigma}\underline{k}_2 r_1\right) + \underline{K}_{v_p+1}\left({}^{u,\varsigma}\underline{k}_2 r_1\right)\right] \end{array}\right] \left[\underline{I}_{v_p}\left({}^{u,\varsigma}\underline{k}_2^* r_1\right) + {}^{u,\varsigma,v}\underline{C}_8^* \underline{K}_{v_p}\left({}^{u,\varsigma}\underline{k}_2^* r_1\right)\right] \right\} \tag{5.112}$$

For canned SRMs that the rotor can loss is overestimated by the imaginary increase of radial length of that can, the active torque generation should be compensated. According to the energy conversion principle, the amount of the overestimated energy should be reversed back to generate torque. Based on the coefficient of correction a in (5.74), the compensated torque is expressed as

$$T_{comp} = (1 - a)P_{can}/(2\pi n) \tag{5.113}$$

where P_{can} is average loss of the rotor can, n is rotor speed. Torque generation now is made up of the three parts as

$$T = -(T_{stator} + T_{statorcan}) + T_{comp} \tag{5.114}$$

5.6 Verification and Discussion

In order to verify the proposed concentric layer structure model and simultaneously study the can effect, the canned induction machine is first analyzed, followed by the canned permanent magnet machine. Further, a canned switched reluctance machine with prominent of double saliency structure and DC excitation is simulated.

5.6.1 Canned Induction Machine

A squirrel cage canned induction machine in Table 5.1 is studied. The number of pole pair $p = 2$ and synchronous speed is 1500 rpm. Compared with ordinary design, the airgap length is enlarged to 1.2 mm to accommodate cans each of which is 0.4 mm in radial length. Meanwhile, the machine is axially shorter to resist eddy current on cans, which means can loss per unit length is reduced.

Magnetic flow due to the application of cans is comparatively presented in Fig. 5.6. The left part is the ordinary machine at rated speed. The ordinary machine

Table 5.1 The canned machine overview

Rated power (kW)	5.1	Bore radius (mm)	38
Rated speed (rpm)	1460	Can alloy	Hastelloy C
Axial length (mm)	42	Outer radius (mm)	54
Airgap width (mm)	1.2	Number of phases	3

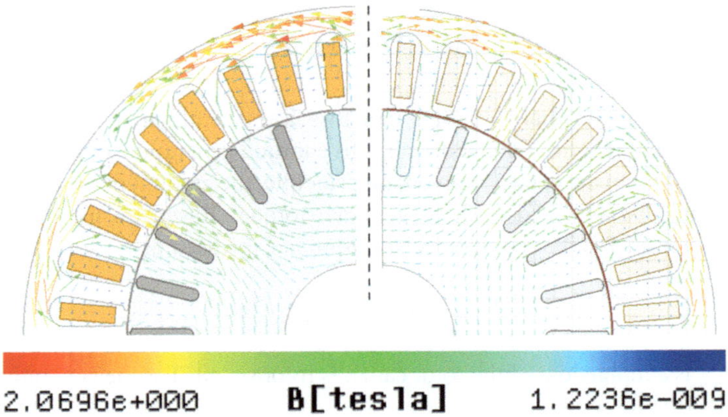

Fig. 5.6 Comparison of flux paths of the ordinary and canned machine under the same condition

has the same geometry as shown in Table 5.1, including the enlarged airgap width, and the only difference is free from cans. The right part is the canned machine under study. It is shown that at rated speed the flux flows on both parts are similar but the density is reduced due to interaction of magnetic field between eddy current on cans and airgap flux.

When operation speed changes, the flux variation due to cans is further investigated. The flux density of stator teeth B_{ST} and yoke and B_{SY} versus rotor speed in terms of slip s_i is shown in Fig. 5.7. The slip is within [0,1]. 0 means the synchronous speed while 1 means stationary. Figure 5.7a shows flux density drops down with s_i increasing, especially on stator poles. As to canned machine in Fig. 5.7b, there is an overall flux drop, due to the electromagnetic interaction on the can shield that helps to alleviate the rate of change of airgap flux.

Iron loss variation whether cans are used is shown in Fig. 5.8 comparatively, with FE simulation as reference. The numerical value is the sum of stator from (5.89), (5.90), (5.91), (5.92), (5.93) and rotor from (5.97). In general, iron loss from both machines drops down with increase of slip. In particular when cans are used, the rate of drop is reduced, due to electromagnetic interaction on cans.

The airgap flux characteristics of the canned machine are further studied. Flux density distribution of both cans under typical operation condition is shown in Fig. 5.9. Transient FE with small time steps is used as reference. Besides the agreement between both methods shown, all sinusoidal waveforms go along the airgap circumference with fluctuation due to higher order flux harmonics. When the

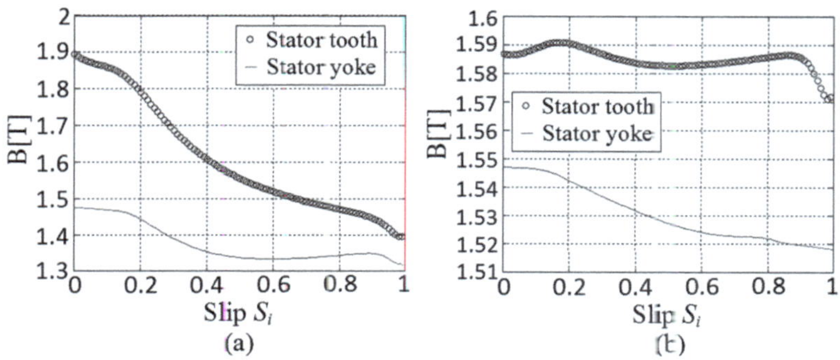

Fig. 5.7 Flux density variation with rotor speed by the proposed model, (**a**) ordinary machine, (**b**) canned machine

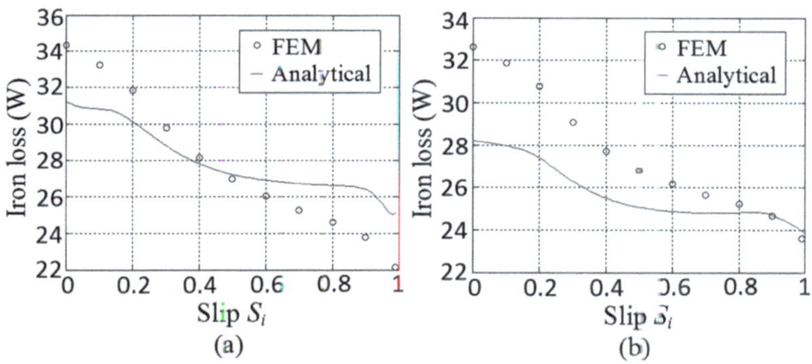

Fig. 5.8 Can effect in terms of iron loss, (**a**) ordinary machine, (**b**) canned machine

relative rotation speed is gradually increasing (synchronous→ rated→ zero), the flux density values are enhanced.

Accordingly, the ordinal flux harmonics are shown in Fig. 5.10. For the stator can layer, the first and third harmonics are prominent, and particularly high under lower speed. In analogy, the lower order harmonics of rotor can flux are the main components and are more prominent when rotor starts. However the amplitude is comparatively much lower, due to lower relative speed between the rotor and phase current.

Loss corresponding to each harmonic is shown in Table 5.2, for which the total loss calculation via FE is applied as reference. The can loss, being much higher compared to iron loss, stems in analogy from eddy current that is approximately in proportion to the square rate of change of flux density. The cans are non-laminated and almost unable to resist eddy current circulation. As to stator can, the loss comes from the first, followed by the third harmonic at each operation condition. The value is remarkably lower at starting, because of the control strategy that only part of

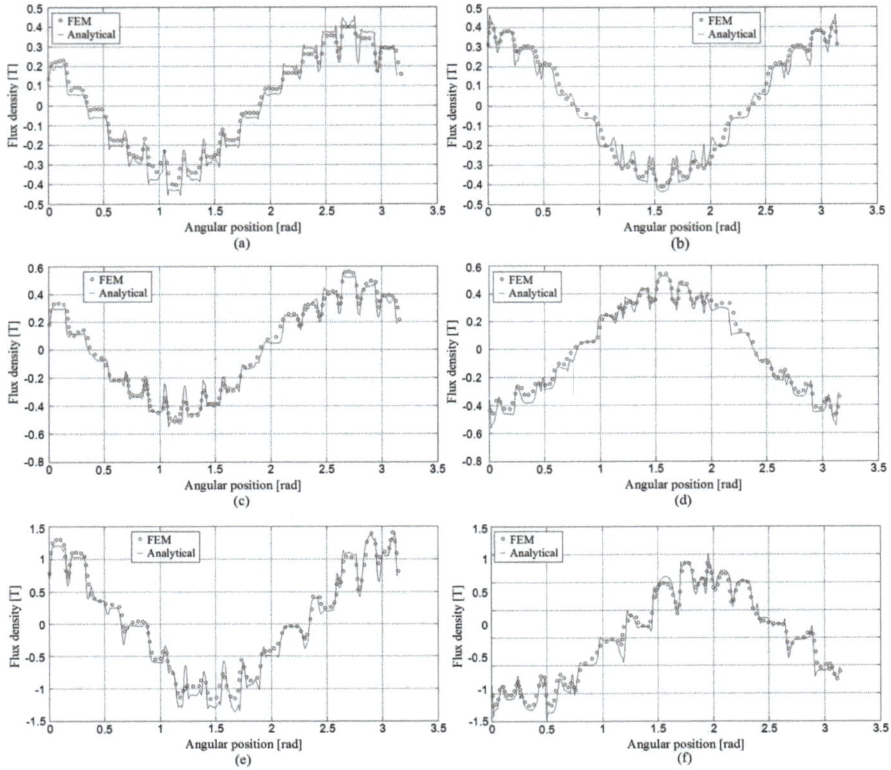

Fig. 5.9 The normal flux density distribution of cans under different operation condition, (**a** and **b**), at synchronous speed, (**c** and **d**), at rated speed, (**e** and **f**), at zero speed, (**a**, **c** and **e**), stator can, (**b**, **d** and **f**) rotor can

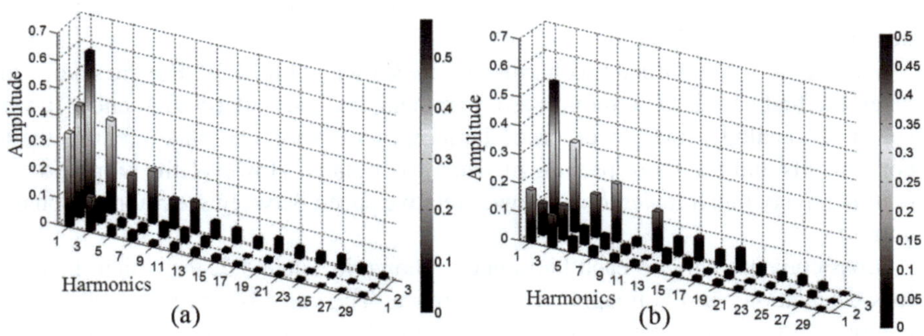

Fig. 5.10 The normal flux density harmonics of cans at different condition, (**a**) stator can, (**b**) rotor can. Note: 1-synchronous speed, 2-rated speed, 3-zero speed

Table 5.2 The can loss (W) from harmonics

Harmonic order	No load		Rated		Zero speed	
	SC	RC	SC	RC	SC	RC
1	585.1	0.0	632.2	0.2	71.1	84.2
3	141.8	1.1	98.3	0.4	3.3	11.1
5	15.9	1.8	8.2	1.9	1.5	1.2
7	15.6	1.2	1.8	1.6	0.6	5.8
9	2.3	0.4	10.2	0.8	0.2	0.1
11	3.5	0.6	7.6	10.5	1.1	1.9
13	0.2	76.9	6.5	85.9	0.2	0.8
15	0.8	59.6	2.1	25.2	0.1	0.6
17	0.1	16.7	0.3	0.1	0.1	0.1
19	0.2	0.2	1.3	0.0	0.3	0.7
21	0.1	0.4	0.4	0.0	0.0	0.0
23	0.0	0.1	0.2	0.0	0.0	0.0
25	0.0	0.0	0.0	0.1	0.3	0.2
27	0.1	0.0	0.0	0.0	0.0	0.0
29	0.0	0.0	0.2	0.0	0.0	0.0
Total	**765.6**	**142.3**	**769.3**	**126.7**	**78.8**	**93.6**
Total (FE)	**801.3**	**167.7**	**808.6**	**155.5**	**84.5**	**104.3**

SC stator can, *RC* rotor can

voltage with lower frequency is used. As to rotor can, loss comes from the fundamental order at zero speed, meaning that the slip is the highest. However when the slip decreases close to 0 at rated or no load operation, loss from low order harmonics is almost negligible. In this occasion the ordinal numbers 13th and 15th serve as working harmonics.

The calculation is compared with measurement. The dSPACE test system is shown in Fig. 5.11. The prototype machine is mechanically connected to a dynamometer via a couple of clutches. Rotor position and speed are recorded by a magnetic encoder, providing 4096 pulses per rotation and one index pulse. Power inverter receives voltage supply up to 400 V. Temperature sensors are used on windings and axial endings of cans to guarantee safe operation. The machine is dry running at low speed and friction loss is not a prime concern. The total loss in consideration includes copper loss, iron loss and can loss, and all are integrated by measuring input and output power. The test includes measurements of electrical and mechanical parameters. The mechanical include shaft speed and torque, while the electrical include input power, voltage and current at different voltage levels.

Figure 5.12 shows the loss varying with increase of excitation current level at no load condition. Both measurement and analytical model agree with each other. It is also shown that the loss increases accordingly with higher excitation. Note that the measured value at any point is higher, because of actual friction and stray loss that is

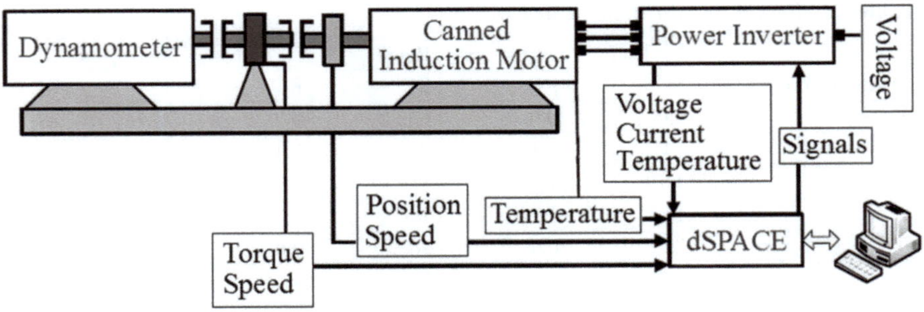

Fig. 5.11 Schematic of the test bench, showing the dSPACE 1103 system

Fig. 5.12 The total loss from the analytical method and experiment

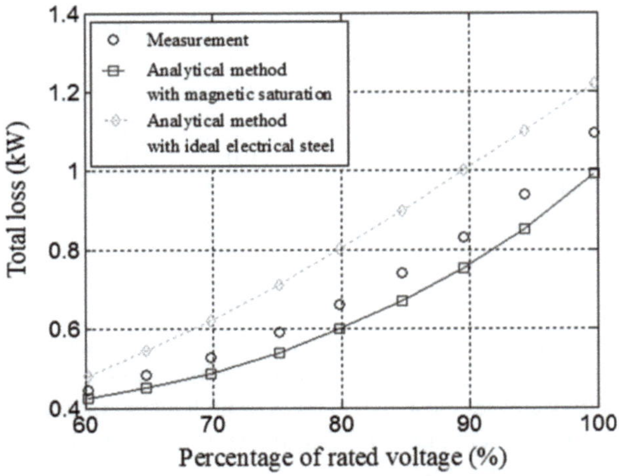

inevitably included in the test system but not taken in the analytical model. In addition, analytically calculated loss without considering magnetic saturation is presented by setting permeability of electrical steel infinite. As a result, loss in value at any point is higher than measurement, indicating the computational gap value. The gap increases with higher voltage percentage, as the airgap flux should have been further resisted by reluctance due to magnetic saturation.

The importance of taking magnetic saturation is further shown in Table 5.3 by comparing numerical values of losses using the actual and ideal electrical steels respectively. Iron loss is a little whereas can loss is much overestimated. Magnetic saturation effect is often neglected or roughly estimated, which is feasible as iron loss is numerically low. However since the can loss is much higher, such simplification may cause inaccuracy.

Table 5.3 Loss (W) by different magnetic materials at no load

	Iron loss		Stator can loss		Rotor can loss	
Rated volt%	I	II	I	II	I	II
60.25	13.1	16.5	301.1	359.3	70.1	97.1
64.79	13.5	18.0	320.0	402.3	74.4	102.1
69.85	14.4	19.6	359.8	448.9	79.8	110.3
75.15	16.1	22.1	406.6	521.0	85.2	120.5
79.84	17.6	24.5	451.4	539.5	92.0	131.1
84.78	19.4	27.0	505.2	659.0	101.5	141.3
89.56	22.0	30.0	575.5	746.3	111.6	151.7
94.34	24.8	33.4	665.2	844.1	125.3	163.1
99.77	28.9	37.8	808.1	998.7	144.4	179.4

I = M1929G, II = Non-saturated ideal steel

5.6.2 *Canned Permanent Magnet Machine*

The PM synchronous machines with tooth concentrated coils have been attracting many interests. Such coils have simple end windings, higher fault tolerance and low cost. Using fractional slot tooth concentrated windings (FSCW) with different poles and teeth were reported [22, 23] for a more efficient design. Especially, $q = [1/3, 1/2]$ presents more desired performances [24, 25]. However, FSCW is inherently with high space harmonic contents. By an appropriate choice of combination of stator slots and pole numbers, a 3-phase 12–10 surface mounted PM (SPM) machine is studied. The model is from [26], but with cans as an update. This section investigates the can effect by updating the model as follows. The machine is canned by enlarging the airgap radial length (Fig. 5.13). The use of tooth concentrated windings provides desired advantages while parasitic effects of harmonics are weakened by a high airgap length. The stator can adhering to stator teeth is to seal slots whereas the rotor can embracing rotor bore is to reduce friction loss and protect PMs. The liquid being pumped is able to go through axially between cans.

Main parameters of the proposed machine are listed in Table 5.4. The axial length of the machine is shorter to reduce can loss by confining eddy current. The airgap radial length is quite high, not only to accommodate cans and PMs, but also to facilitate liquid to go through. Thickness value of each can is 0.5 mm, a compromise between can loss generation and mechanical stiffness. As to the can material, Hastelloy C has higher electrical resistivity than stainless steel (Austenite). The PMs are cobalt based with higher temperature tolerance, an essential feature when the can loss value is remarkably high. The temperature rise has a direct impact on output. The electrical resistivity of cans rises up with temperature, resulting in higher can loss that in turn affects PMs. For simplicity, constant steady-state working temperature value is applied throughout the modeling. The value is predetermined by electromagnetic thermal coupled analysis via FE.

Flux distribution with and without cans is comparatively shown in Fig. 5.14. Due to the very large airgap, the peak flux density is relatively low and the machine is unlikely to saturate. In Fig. 5.14, an ordinary machine in (a) is updated by simply

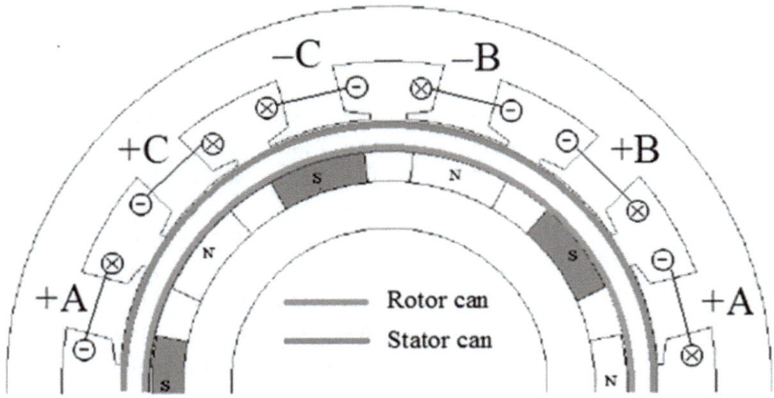

Fig. 5.13 Schematic diagram of a canned SPM machine

Table 5.4 The studied 12–10 canned SPM machine parameter overview

Rated power (kW)	3.2	Rotor can thickness (mm)	0.5
Rated voltage (V)	220		
Can alloy	Hastelloy C/stainless	Stator can thickness (mm)	0.5
Axial length (mm)	62	Rated speed (rpm)	3000
Outer radius (mm)	81.5	Magnet material	$SmCo_{18}$
Airgap width with magnets (mm)	7.4	Thickness of magnets (mm)	5

inserting cans in airgap in (b). The flux flows on both sides are similar but the density is reduced on the canned machine. This is due to interaction of electromagnetic field on cans to consume magnetic energy that not only alleviate the rate of change, but reduces the peak of airgap flux.

The solution is quite linear due to absence of magnetic saturation, and the airgap field can be obtained by the superposition from both PMs and stator current. The airgap flux density harmonics of the ordinary machine without cans are respectively shown in Fig. 5.15. In any of the cases, the ordinal number 5 is the working harmonic and there is only a slight difference on amplitude values whether armature windings are excited. This is mainly because of the enlarged airgap and current flow in q-axis. Further when cans are inserted, both the working and high order harmonics are reduced due to additional energy consumption on cans.

The can effect is illustrated by investigating airgap flux density waveforms at a particular rotor position. In Fig. 5.16 besides agreement between FE and the analytical method, the waveform represents all PM sheets on the rotor bore. When cans are inserted in (b), the overall tendency of flux variation remains. However due to eddy current on cans, the waveform peak is reduced while the local spikes and sinks are suppressed.

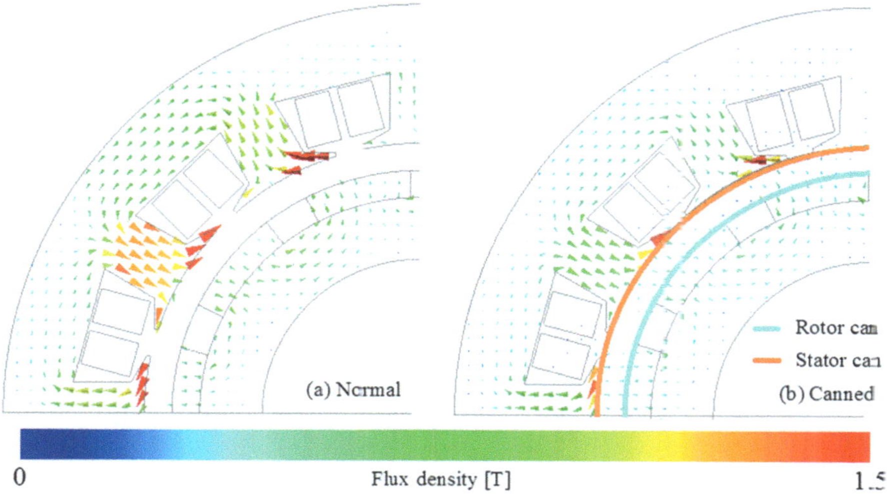

Fig. 5.14 Flux path and density distribution due to cans, (**a**) ordinary machine, (**b**) canned machine

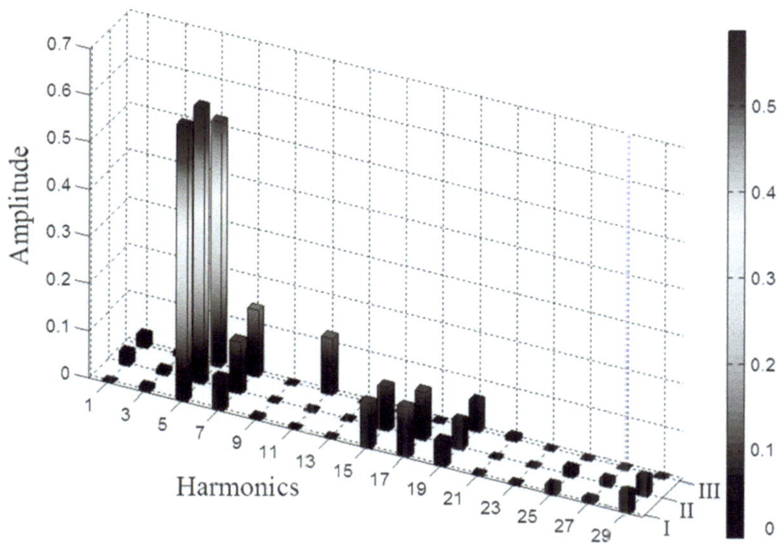

Fig. 5.15 Airgap flux density harmonics at different occasions of the studied machine, no load: (*I*) PMs excited only without cans, (*II*) both PMs and armature coils excited without cans, (*III*) both PMs and coils are excited when cans are inserted

The airgap flux density harmonics of the canned machine at typical operation condition is shown in Fig. 5.17. The fifth order still serves as the main working harmonic and its amplitudes at rated condition is slightly higher than that of no load

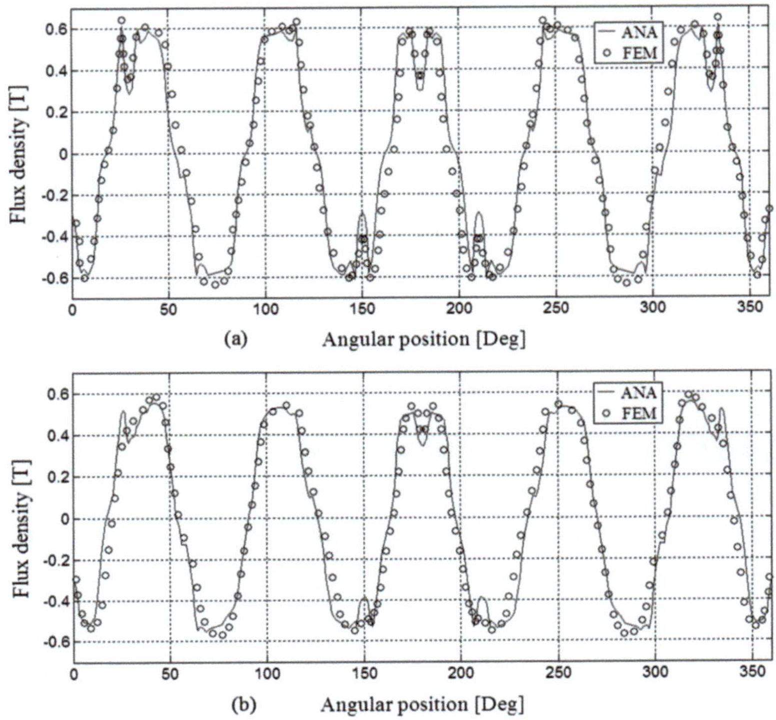

Fig. 5.16 Radial airgap flux density distribution, (**a**) without cans, (**b**) with cans

but is reduced sharply at zero rotor speed. The amplitude of high order harmonics increases especially at zero speed, causing more loss.

The can loss values from each harmonic at typical operating condition are presented in Table 5.5. Besides agreement between FE and analytical calculations, it is shown that the operation condition has great impact on the can loss and there exists quite a difference upon stator or rotor can. As to the total loss, the rated is slightly higher than no load while the zero-speed drops down sharply. As to the stator can loss, the fifth order is the main component accounting for 96.9% and 94.4% respectively. Specifically, loss at rated speed is 5–10% higher than no load due to a higher working harmonic. The rotor can loss is quite lower, owning to synchronous speed. However at the point when rotor starts, higher rotor can loss occurs. In that case, losses from both cans are similar as both are imposed in the same airgap flux variation.

Besides the operation condition, geometry and material of can shield determine loss generation. Due to the requirements of mechanical stiffness reasons, the can radial length is expected higher. The material should be non-magnetic, higher electrical resistance, higher thermal conductance and ease of tooling and assembly. Up to now, the widely adopted are nickel based Hastelloy C, as well as conventional stainless steel.

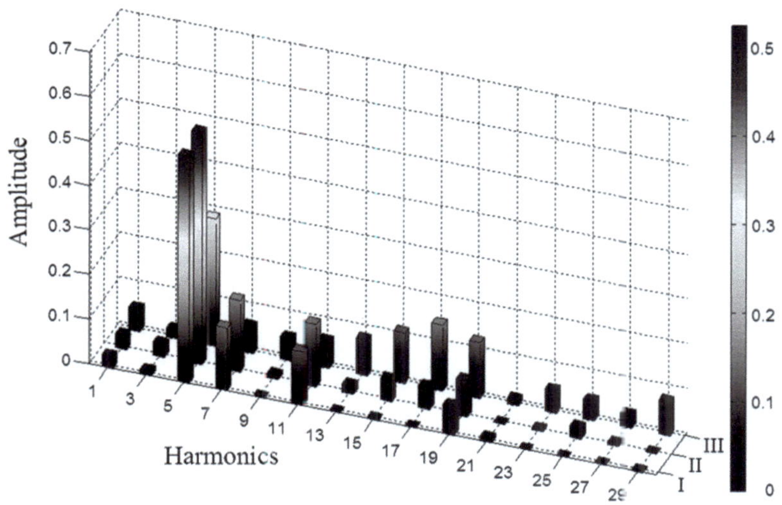

Fig. 5.17 Airgap flux density harmonics of the canned machine at different operation condition, (*I*) no load, (*II*) rated, (*III*) start

Table 5.5 The can loss generation (W) from harmonics by the proposed analytical model

	No load		Rated		Zero speed	
Harmonic order	SC	RC	SC	RC	SC	RC
1	4.4	0.0	1.9	0.0	0.3	0.4
3	1.8	0.1	0.1	0.0	0.1	0.1
5	262.9	0.2	291.2	0.4	29.8	32.8
7	0.6	0.0	1.8	0.6	0.1	0.3
9	0.3	0.1	0.1	0.3	0.2	0.2
11	0.5	0.0	1.6	0.5	0.1	0.2
13	0.2	0.0	0.1	0.1	0.0	0.1
15	0.1	0.1	0.1	0.2	0.0	0.0
17	0.1	6.3	0.3	7.4	0.4	0.9
19	0.2	0.1	10.6	0.7	0.5	0.2
21	0.1	0.0	0.4	0.0	0.0	0.0
23	0.0	0.0	0.2	0.0	0.0	0.2
25	0.0	0.0	0.0	0.1	0.0	0.2
27	0.1	0.0	0.0	0.0	0.0	0.0
29	0.0	0.0	0.2	0.0	0.0	0.0
Total	**271.3**	**3.7**	**308.6**	**10.5**	**31.5**	**35.6**
Total FEM	**260.0**	**1.2**	**282.0**	**15.3**	**29.9**	**37.7**

SC stator can (0.5 mm), *RC* rotor can (0.5 mm)

Figure 5.18 shows the stator can loss varying with radial length and material at no load, rated and start respectively. Again both FE and proposed analytical methods agree. For any of the three operation condition, the magnetic saturation is not a

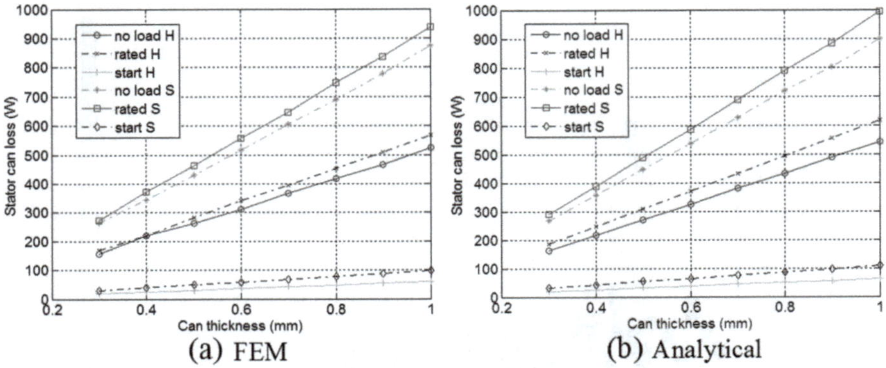

Fig. 5.18 The stator can loss variation as a function of the can radial length and material at typical operation conditions, (**a**) FEM, (**b**) analytical. In either case, *H* hastelloy, *S* stainless steel

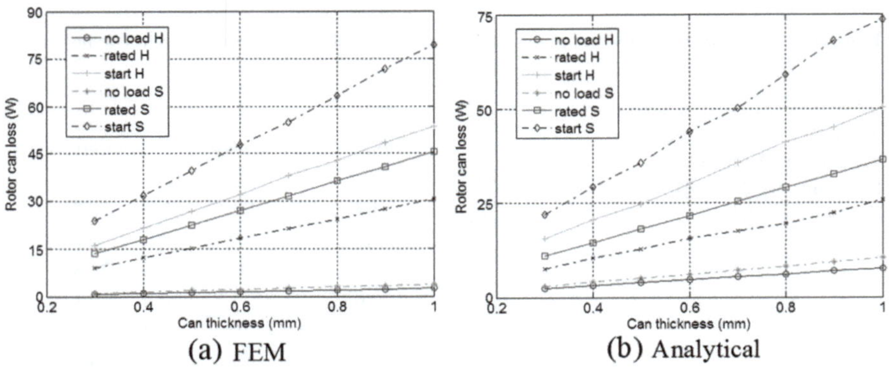

Fig. 5.19 The rotor can loss variation as a function of radial can thickness value and material at typical operation conditions, (**a**) FEM, (**b**) analytical. In either case, *H* hastelloy, *S* stainless steel

determinant factor due to large airgap. Loss of the Hastelloy of higher electrical resistance is lower than that of stainless steel. Loss increases linearly with either rotor speed or can thickness. As to the rotor can loss in Fig. 5.19, besides the above assumptions, the loss is much lower due that the rotor can moves with synchronous speed.

The can loss varying with rotor speed is shown in Fig. 5.20 with specifications in Table 5.4. The proposed analytical method is applied within full range of speed and compared with FE at typical points. The analytical method can yield accurate results. For one thing, it is feasible to neglect nonlinearity due to large airgap length that disables magnetic saturation. For the other, modeling of MMFs and Poynting's energy flow method is accurate. The proposed method facilitates the design process of canned machines with significant computation savings.

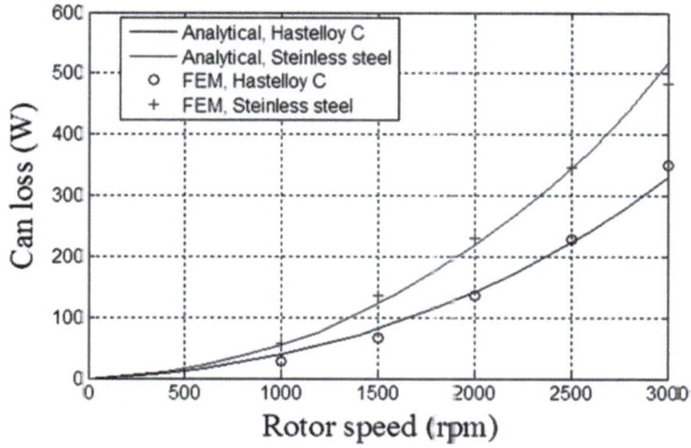

Fig. 5.20 The can loss variation with rotor speed

Table 5.6 The simulated canned switched reluctance machine

Rated power (kW)	20	Rotor can thickness (mm)	0.3
Rated voltage (V)	540		
Can alloy material	Hastelloy C/stainless	Stator can thickness (mm)	0.3
Axial length (mm)	125	Rated speed (rpm)	6000
Stator bore radius (mm)	67.5	Rotor bore radius (mm)	33.95
Airgap width with cans (mm)	0.9	Hysteresis phase current (A)	170

5.6.3 Canned Switched Reluctance Machine

Specifications of a 3-phase 12/8 canned SRM under study in this chapter are shown in Table 5.6. Both cans are 0.3 mm in radial length, the lowest value that helps to reduce the can loss. The loss, including that from outer can (OC, or stator can) and inner can (IC, or rotor can), are analytically calculated (ANA) and finite element method (FEM) is used to compare. Figure 5.21 shows the agreement of both results and the value rises up geometrically with rotating speed. It is also shown that the can loss value is remarkably high, due to the following reasons. (1) the relatively high axial length that enhances eddy current flow on cans; (2) less number of turns per pole ($N_p = 10$) with high excitation 170A that leads to higher rate of change of phase current especially at the progress of phase defluxing; (3) Unlike AC machines, the rotor can shield also contributes considerable loss; (4) The control of fixed on-off angles also leads to sharp increase of can loss.

Figure 5.22 shows the loss of rotor with speed whether cans are added. In analogy, the waveform varies in a similar way because of the same generation principle. However the value is much lower comparatively due to lamination that helps to resist eddy current. Note that the value drops down slightly if cans are used. This is because when cans are imposed in the airgap field, the eddy current generated tends to couple with that field, which helps to reduce loss by resisting the degree of flux fluctuation.

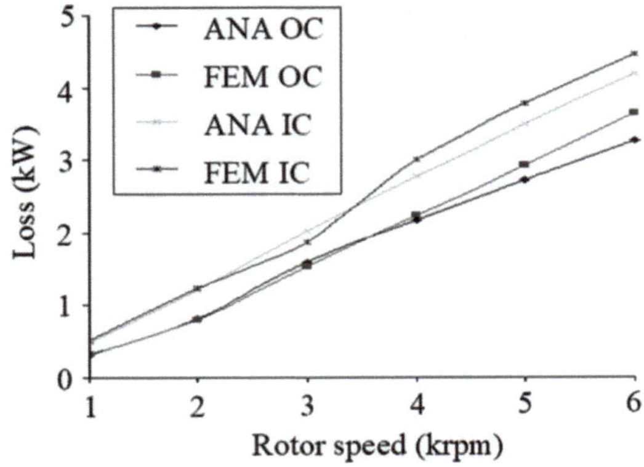

Fig. 5.21 The can loss variation with rotor speed

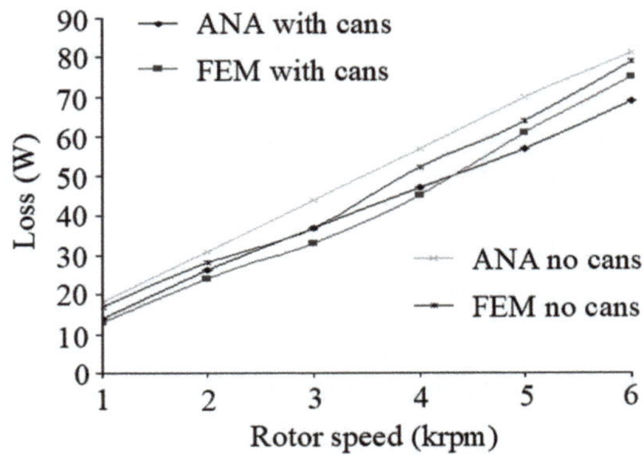

Fig. 5.22 The rotor loss variation with rotor speed

The torque-speed relationship whether cans are present is shown in Fig. 5.23. Results show the consistency of both methods. Generally, the on-average torque value drops down when rotor speed is increasing. When cans are inserted, the active torque drops down due to energy consumption by the high can loss generation.

In Chap. 3, it is known that the can loss mainly occurs at the rotor position range corresponding to phase defluxing progress where phase current is undergoing the drop-down. Here the range is defined as a sector of rotor position duration measured in mechanical degree. Figure 5.24a shows loss and torque variation as a function of the duration variation. If the defluxing process maintains longer duration, the average loss of each can decrease accordingly. This is because of the reduced rate of change of airgap flux. Meanwhile, the average torque is reduced (Fig. 5.24b), due to the back electro-magnetic force (EMF) by the longer duration of phase defluxing.

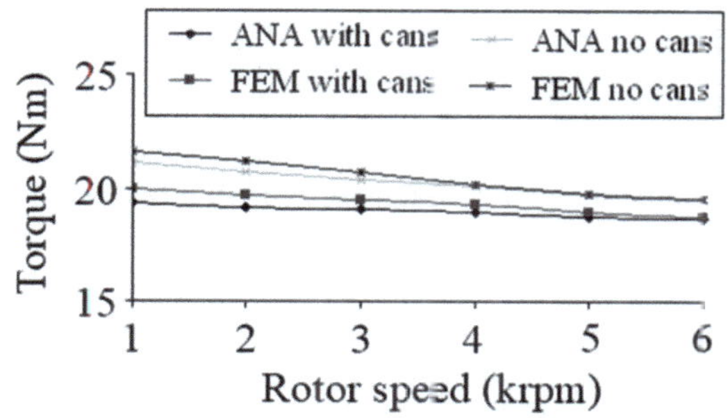

Fig. 5.23 The torque-speed relationship with/without cans

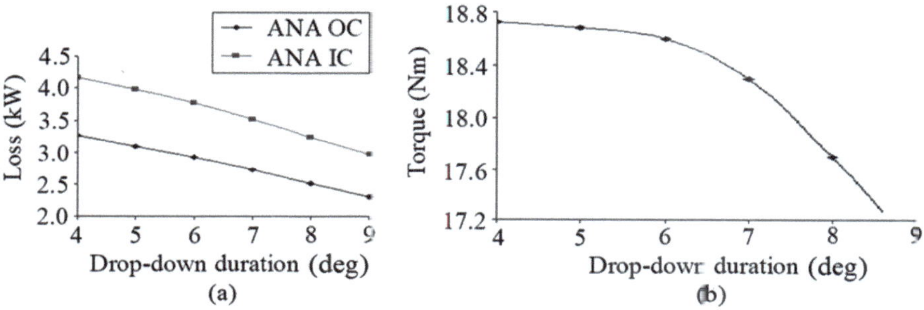

Fig. 5.24 The torque and loss values under different durations of phase defluxing

5.7 Chapter Summary

In this chapter, an analytical model of concentric layer structure under cylindrical coordinate in 3 dimensions is introduced to calculate electrical machines and the can effect, based on spatial distribution of magneto motive force (MMF) in Chap. 4. First, the electromagnetic field for each layer is formulated in terms of the magnetic vector potential under phase current harmonics of Fourier discretization. According to the field distribution, all quantities of interest with regard to the machine performance are derived by applying Maxwell's equations. Loss of each layer is obtained by applying Poynting's theorem of energy conversion, while torque is obtained via Maxwell stress tensor. This model is in particular convenient to analyze the can effect and at the same time to compare with the ordinary machine by simply adding more layers in airgap as can shields.

In addition, the saliency effect, magnetic saturation, MMF due to rotor magnets, the axial end effect, the can shield effect, as well as the magnetic field due to the current potential component of DC excitation, are modeled. Each consideration can be conveniently incorporated as add-ons to account for characteristic features of the machine being studied. Note that the double saliency, such as SRM, leads to time varying radial length of airgap that is modeled by setting the airgap layer constant while using a variable radial length of an inner layer, such as the rotor can layer, with an energy compensation principle. Such assumption also holds true when magnetic saturation is taken. By letting the equivalent airgap radial length time dependent, and permeability of iron be deemed infinite against that of air, a variable radial length of an inner layer is likewise applied. The proposed model facilitates flexible design by evaluating electromagnetic field in a fast and efficient approach.

Using this model, the can effect of a canned induction machine is first analyzed. By comparatively studing both the ordinary and canned machines, the difference in electromagnetic field is shown, followed by ohmic losses. In particular, both rotor and stator can losses at typical condition are intensively calculated, simultaneously with FE and experiment as verification. It is shown that the stator can loss is the highest and increases sharply with rotor speed and phase current, in which the fundamental and the third flux harmonic orders account for almost losses. The rotor can loss is comparatively lower due to the slip. For the rotor can, similarly the first and third harmonic takes principle effect close to zero speed while higher orders will work at rated or close to no load condition. By comparing can loss respectively under the ideal and practically saturated electrical steels, it is stressed the importance of finely taking magnetic saturation.

Then for PM machines, a 12–10 SPM prototype with concentrated windings, which attracts great research interest, is canned and studied. The airgap flux and magnetic field with cans show that effective airgap flux density is reduced meanwhile higher order harmonics are suppressed. The difference of losses on each can shows that the stator can loss constitutes the main part where the fifth order harmonic is prominent while the rotor can loss is much lower due to synchronous speed. Further, loss variation in terms of can radial length and material shows that a thinner wall of higher electrical resistance helps reduce loss, especially at high rotor speed.

Lastly, a classical 3-phase 12/8 DC switched reluctance machine is canned. Compared with AC machines, the loss value is much higher. This is because DC excitation brings much more current harmonics and the rotor can contributes a considerable part of loss as well. Further it is shown that the phase defluxing process featured by the rate of drop of phase current has great influence on torque and can loss generation. With longer drop-down duration, both the loss and torque values are reduced.

References

1. B. Vaseghi, F.M. Tabar, Modeling of non-salient PM synchronous machines under stator winding inter-turn fault condition: dynamic model – FEM model, Conference of Vehicle Power & Propulsion, Arlington, TX, in *USA*, (2007)
2. A.M. El-Serafi, X. Liang, S.O. Faried, Factors affecting accuracy of determining the saturated synchronous machines cross-magnetizing parameters by the finite element method, International Conference on Electrical Machines, ICEM 2008, Vilamoura, Portugal

3. N. Sadowski, M.L. Mazenc, Evaluation and analysis of iron losses in electrical machines using the rain-flow method. IEEE Trans. Magn. **36**(4), 1923–1926 (2000)
4. Q. Yu, B. Bilgin, A. Emadi, Loss and efficiency analysis of a switched reluctance machine for traction application with a new calculation method. IEEE Trans. Ind. Electron. **5**(62), 3072–3080 (2015)
5. G. Dajaku, D. Gerling, Stator slotting effect on magnetic field distribution of salient pole synchronous permanent machines. IEEE Trans. Magn. **46**(9), 3676–3683 (2013)
6. T. Lubin, S. Mezani, A. Rezzoug, Exact analytical method for magnetic field computation in the air gap of cylindrical electrical machines considering slotting effects. IEEE Trans. Magn. **46**(4), 1092–1099 (2010)
7. F. Dubas, C. Espanet, Analytical solution of the magnetic field in permanent-magnet motor taking into account slotting effect. IEEE Trans. Magn. **45**(5), 2097–2109 (2009)
8. Q. Yu, D. Gerling, Analytical modeling of a canned switched reluctance machine with multilayer structure. IEEE Trans. Magn. **49**(9), 5069–5082 (2013)
9. Z.Q. Zhu, D. Howe, Instantaneous magnetic field distribution in pm brushless dc motors – Part III: effect of stator slotting. IEEE Trans. Magn. **29**(1), 143–151 (1993)
10. M. Ojaghi, S. Nasiri, Modeling of Eccentric squirrel cage induction motors with slotting effect and saturable teeth reluctances. IEEE Trans. Energy Convers. **29**(3), 619–627 (2014)
11. D. Zarko, D. Ban, T.A. Lipo, Analytical calculation of magnetic field distribution in slotted air gap of a surface permanent-magnet motor using complex relative air-gap permeance. IEEE Trans. Magn. **42**(7), 1828–1837 (2006)
12. Q. Yu, X.S. Wang, Y.H. Cheng, Electromagnetic modeling and analysis of can effect of a canned induction electrical machine. IEEE Trans. Energy Convers. **31**(4), 1471–1428 (2016)
13. Q. Yu, X.S. Wang, Y.H. Cheng, Electromagnetic calculation and characteristic analysis of can effect of a canned permanent magnet motor. IEEE Trans. Magn. **52**(2), 1 (2016)
14. O. Laldin, S.D. Sudhoff, S. Pekanrek, Modified Carter's coefficient. IEEE Trans. Energy Convers. **30**(3), 1133–1134 (2015)
15. D. Gerling, G. Dajaku, Three-dimensional analytical calculation of induction machines with multilayer rotor structure in cylindrical coordinates. Electr. Eng. **86**(4), 199–211 (2004). Archiv fuer Elektrotechnik, Springer
16. G. Dajaku, D. Gerling, Airgap flux density characteristics of salient pole synchronous permanent magnet machines. IEEE Trans. Magn. **48**(7), 2196–2204 (2012)
17. D. Gerling, Three dimensional analytical calculation of the permanent magnet motor in cylindrical coordinates, Technical report, Professur fuer Antrienbstechnik und Automation, University of Bumdeswehr Muenchen, Munich, Germany, 2003
18. G. Bertotti, General properties of power losses in soft ferromagnetic material. IEEE Trans. Magn. **24**, 621–630 (1988)
19. H. Mosebach, Eisenverluste von NK Super E-Core Materialen unterschiedlicher dicke und vergleich mit Klassischen Blechqualitaeten, 1999, Availabe at: http://www.iem.ing.tu-bs.de/paper/1999/moseb_99.htm
20. F. Deng, An improved iron loss estimation for permanent magnet brushless machines. IEEE Trans. Energy Convers. **14**(4), 1391–1395 (1999)
21. Y. Chin, J. Soulrad, Modelling of iron losses in permanent magnet synchronous motors with field-weakening capability for electric vehicles, International Battery, Hybrid and Fuelcell Electric Vehicle Symposium & Exhibition, EVS 19., Korea, 2002
22. J. Wang, K. Atallah, Z.Q. Zhu, D. Howe, Modular three phase permanent magnet brushless machines for in-wheel applications. IEEE Trans. Veh. Technol. **57**(5), 2714–2720 (2008)
23. J. Li, D.W. Choi, D.H. Son, Y.H. Cho, Effects of MMF harmonics on rotor eddy current losses for inner rotor fractional slot axial flux permanent magnet synchronous machines. IEEE Trans. Magn. **48**(2), 839–842 (2012)
24. G. Dajaku, W. Xie, D. Gerling, Reduction of low space harmonics for the fractional slot concentrated windings using a novel stator design. IEEE Trans. Magn. **50**(5), 99–110 (2006)

25. D. Ishak, Z.Q. Zhu, D. Howe, Comparison of PM brushless motors, having either all teeth or alternate teeth wound. IEEE Trans. Energy Convers. **21**(1), 95–103 (2006)
26. G. Dajaku, D. Gerling, *Eddy current loss minimization in rotor magnets of PM machines using high efficiency 12-teeth 10-slots winding topology, International Conference on Electrical Machines and Systems* (ICEMS, China, 2011)

Chapter 6
Thermal Analysis of a Canned SRM

6.1 Background

The application of can shields makes electrical machines thoroughly different in electromagnetic performance. Thermal analysis is particularly necessary [1–4], as the can loss value is considerably higher than copper or iron loss. Identification of temperature rise on the cans plays the central role, including the overall rise and local thermal sensitive regions. A fast and accurate mathematic heat transfer model is essential and one classic is the lumped parameter network [5–7]. Derived from the machine geometry, heat sources, materials and cooling, this model takes all components and heat transfer mechanisms [8–10]. However, the heat flow is considered constant within the volume being estimated, whereas in fact it alters [11, 12], causing systematic mistake.

The necessity of using compensation elements as a measure of calculation accuracy improvement is illustrated in previous work [13, 14]. Temperature distribution, lumped network model and comparative simulations of a block component as example are shown in Fig. 6.1. Assuming that heat source is evenly distributed in the block and lateral 4 faces of the block is thermally isolated, heat only flows along the length of the block. Temperature distribution along the length is shown in Fig. 6.1a. Accordingly calculation accuracy is compared by FE and the traditional network (Fig. 6.1c), showing that the network overestimates the value especially at the mid-point. The improvement is made by using additional compensation elements t_{comp} into the network model that helps to correct the numerical mistake, splitting the block into a couple of halves.

The compensation elements are obtained by

$$t_{comp1} = (P/4)(R_{TH,1}/2) \tag{6.1}$$

$$t_{comp2} = (P/4)(R_{TH,2}/2) \tag{6.2}$$

$$t_{comp} = t_{comp1} = t_{comp2} \tag{6.3}$$

Q. Yu et al., *Analysis and Mathematical Models of Canned Electrical Machine Drives*, https://doi.org/10.1007/978-981-13-2745-2_6

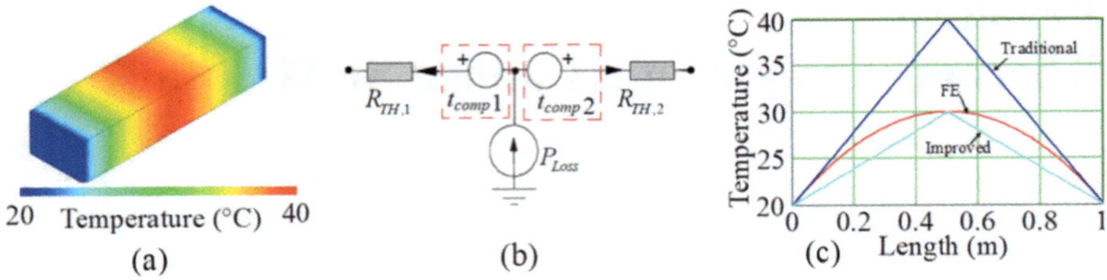

Fig. 6.1 Introducing compensation elements t_{comp} into the traditional lumped parameter thermal network, FE is used as reference, (**a**) Temperature distribution by FE analysis, (**b**) The lumped parameter network model with compensation elements, (**c**) Numerical waveforms and comparison

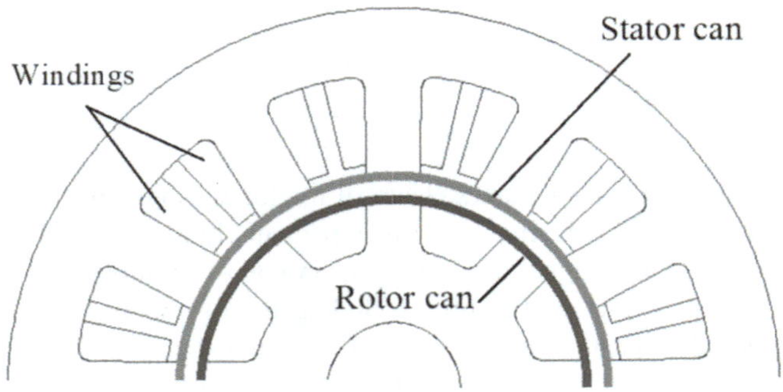

Fig. 6.2 Geometrical overview of the studied canned machine

Generally, the compensation t_{comp} equals to heat flow by half of the source that is $P/2$, multiplied with resistance of either half $R/2$. All calculations are shown in Fig. 6.1c. The traditional network overestimates temperature rise especially at the midpoint of the block while the improved network with t_{comp} and FE fit each other. Such a mistake of the traditional network is previously eliminated by curving fitting procedure. Such a proposal is simple, but decisive. For complicated systems of more than one component, t_{comp} is introduced into the network.

This chapter presents the thermal characteristics of canned electrical machines, with particular attention on calculation accuracy improvement by compensation elements based on the traditional network model. The 3-phase 12/8 canned switched reluctance machine in Fig. 6.2 is studied, and specifications are listed in Table 5.5 of Chap. 5. The characteristic features are that due to the vast source of ohmic can loss, consequent temperature rise makes a difference, which leads to a sharp temperature jump when the machine is running. Further a thermal-electro coupled phenomenon occurs, due to variation of thermal and electrical conductivity of the can material arising form that jump that affects both eddy current heat transfer. Organization of this chapter is as follows. In Sect. 6.2, the network model is systematically introduced, including the deduction of heat conduction, convection, contact based on the studied SRM with particular attention on modeling of cans. Then deduction of compensation elements and modeling of armature coils are introduced. In Sect. 6.3,

Table 6.1 List of thermal resistances

Thermal resistance R_{th}	Between (and)
$R_{th\_fr\_am}$	Frame and ambient
$R_{th\_SY\_fr}$	Stator yoke and frame
$R_{th\_SY\_ST}$	Stator yoke and stator teeth
$R_{th\_wi\_ST}$	Windings and stator teeth
$R_{th\_wi\_SY}$	Windings and stator yoke
$R_{th\_wi\_EW}$	Windings and end windings
$R_{th\_ST\_OC}$	Stator teeth and outer can (stator can)
$R_{th\_OC\_IC}$	Outer can and inner can (rotor can)
$R_{th\_IC\_RT}$	Inner can and rotor teeth
$R_{th\_RT\_RY}$	Rotor teeth and rotor yoke
$R_{th\_RY\_sh}$	Rotor yoke and shaft
$R_{th\_wi\_SA}$	Windings and slot air
$R_{th\_OC\_SA}$	Outer can and slot air

Table 6.2 List of compensation elements

Compensation element t	Towards (→)
$t_{ST \to OC}$	Stator teeth → outer can
$t_{OC \to ST}$	Outer can → Stator teeth
$t_{OC \to AG}$	Outer can → air gap
$t_{IC \to AG}$	Inner can → air gap
$t_{IC \to RT}$	Inner can → rotor teeth
$t_{RT \to IC}$	Rotor teeth → inner can
$t_{RT \to RY}$	Rotor teeth → rotor yoke
$t_{RY \to RT}$	Rotor yoke → rotor teeth
$t_{RY \to sh}$	Rotor yoke → shaft
$t_{EW \to wi}$	End windings → windings
$t_{wi \to EW}$	Windings→ end windings
$t_{wi \to ST}$	Windings → Stator teeth
$t_{ST \to wi}$	Stator teeth → windings
$t_{SY \to ST}$	Stator yoke → Stator teeth
$t_{ST \to SY}$	Stator teeth → Stator yoke
$t_{SY \to wi}$	Stator yoke → windings
$t_{wi \to SY}$	Windings → Stator yoke
$t_{SY \to fr}$	Stator yoke → frame

the developed model is discussed by comparing with FE and measurements. Meanwhile, features of temperature distribution on cans are discussed. In Sect. 6.4, the thermal- electro coupling effect on cans is discussed.

As one prerequisite, definition of main parameters in the proposed network model of this chapter is listed in Tables 6.1, 6.2, 6.3 and 6.4, including thermal resistances R_{th}, compensation elements t, heat sources P and thermal conductivity k_c.

Table 6.3 List of heat sources

Heat source P	Component
P_{SY}	Stator yoke
P_{ST}	Stator teeth
P_{wi}	Windings
P_{EW}	End windings
P_{OC}	Outer can
P_{IC}	Inner can
P_{RT}	Rotor teeth
P_{RY}	Rotor yoke

Table 6.4 List of thermal conductivities

Thermal conductivity k_c	Component
k_{c_aly}	Alloy (machine body)
k_{c_cop}	Copper
k_{c_wi}	Windings (with coating)
k_{c_can}	Can
k_{c_air}	Air
k_{c_liq}	Liquid (stationary water)
$k_{c\_eff\_liq}$	Moving liquid

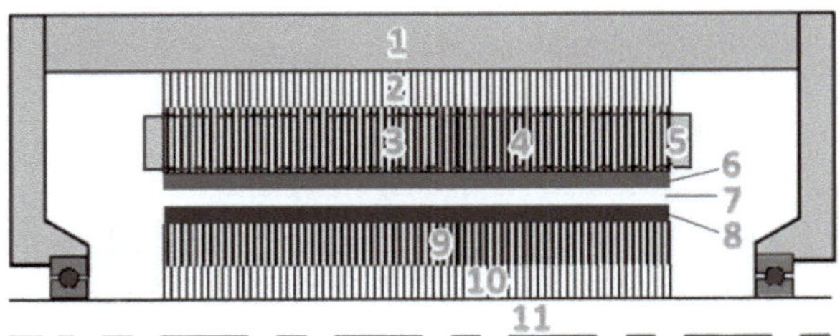

Fig. 6.3 Discretizing the machine into connected components

6.2 The Thermal Network Model

6.2.1 Model Overview

In Fig. 6.3, the canned SRM is divided into 11 components, namely (1) frame (abbreviated as *fr*), (2) stator teeth (*ST*), (3) stator yoke (*SY*), (6) outer can or stator can (*OC*), (7) airgap (*AG*), (8) inner can or rotor can (*IC*), (9) rotor teeth (*RT*), (10) rotor yoke (*RY*), (11) shaft (*sh*), and coils. A coil is further divided due to different cooling condition into (4) windings (*wi*) and (5) end windings (*EW*). The nodal temperature T of each component is defined as T_{fr}, T_{wi}, T_{IC}, T_{SY}, T_{EW}, T_{RT}, T_{ST}, T_{OC} and T_{RY}. Note that abbreviated subscript indicates component name. According to Fig. 6.1, T stands for the average value of the regarded component.

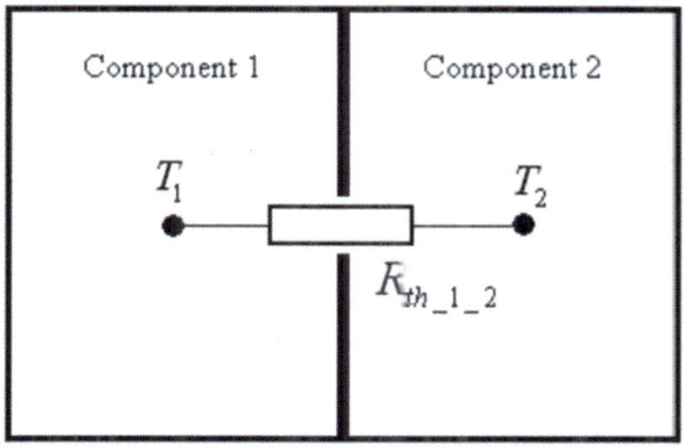

Fig. 6.4 A calculation example of thermal resistance

6.2.2 *Heat Sources*

Heat source P from each component is labeled using the same principle. Eight out of the eleven defined components are carrying sources. Copper loss includes P_{wi} and P_{EW}, iron loss includes P_{ST}, P_{SY}, P_{RT} and P_{RY}, and can loss includes P_{OC} and P_{IC}. The loss values are obtained by a magnetic model of multilayer structure in Chap. 5. The energy flow of each component is calculated to get loss values by Poynting's theorem.

6.2.3 *Thermal Resistances*

The naming principle of the principal resistances R_{th} in subscript is similar by using abbreviated names for the couple of connected components following subscript "*th*". Note that the connection is interchangeable, e.g., the resistance between stator teeth and stator yoke can be also written as $R_{th\_ST\_SY} = R_{th\_SY\_ST}$. As an example for a couple of connected components 1 and 2 (Fig. 6.4), the resistance $R_{th\_1\_2}$ between the midpoint T_1 and T_2 is written as

$$R_{th\_1\_2} = R_{th\_2\_1} = \frac{1}{2}(R_{th\_1} + R_{th\_2}) \tag{6.4}$$

Heat transfer includes conduction, convection and contact, and each type can be seen as such a couple of connected components. Note that in some occasions such as convection, one of the components is not solid but serves as a means of cooling. In the following, each type of heat transfer is studied.

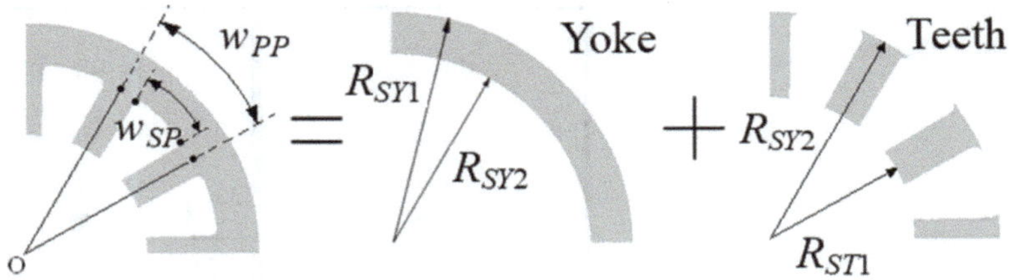

Fig. 6.5 Connecting resistances between stator teeth and yoke

6.2.3.1 Conduction

$R_{th\_ST\_SY}$, $R_{th\_ST\_OC}$, $R_{th\_IC\_RT}$, $R_{th\_RT\_RY}$ and $R_{th\_wi\_EW}$ fall into this category. Components, like stator yoke and rotor yoke R_{th_SY} and R_{th_RY}, both cans R_{th_IC} and R_{th_OC} as well as frame R_{th_fr}, are seen as hollow cylinders that thermal resistance can be easily obtained. For example, thermal resistance of the cylindrical stator yoke is written as

$$R_{th\_SY} = \frac{\ln\left(R_{SY1}/R_{SY2}\right)}{2\pi k_c L_z \lambda} \tag{6.5}$$

where R_{SY1} and R_{SY2} are outside and inner side radius of the cylinder respectively, k_c is the thermal conductivity, L_z is axial length and λ is the packing factor concerning lamination. As an extended step, the stator/rotor poles as a group R_{th_ST} or R_{th_RT} are seen as one cylinder as well, as long as the slot cavities between poles are deemed as a heterogeneous sector that hinders heat flow. Likewise according to (6.5), thermal resistance of stator poles is written as

$$R_{th\_SY} = \frac{\ln\left(R_{SY2}/R_{ST1}\right)}{2\pi k_c L_z \lambda_1} \tag{6.6}$$

where R_{ST1} is inner side of stator teeth (inner bore of stator). Compared with (6.5), improvements fall into λ_1, which donates the approximation factor that donates the percentage of teeth volume against the total teeth plus the slot cavity volume.

Figure 6.5 further illustrates the connecting resistance between stator poles and yoke. Due to slot cavities that hinder heat flow at a fixed rate, a geometrical coefficient $c_{CR\_ST\leftrightarrow SY}$ is defined to account for the percent of iron that the heat flow passes through. Note that the subscript "*CR*" is short for "conduction/contact ratio", and "↔" indicates being unchangeable, that is $C_{CR\_SY\leftrightarrow ST} \neq C_{CR\_ST\leftrightarrow SY}$.

$$c_{CR\_ST\leftrightarrow SY} = \frac{w_{PP} - w_{SP}}{w_{PP}} \tag{6.7}$$

where w_{PP} and w_{SP} are respectively pole pitch and slot pitch measured from circumferential direction respectively (Fig. 6.5). Then the resistance between stator teeth and yoke is written as

$$R_{th\_ST\_SY} = \frac{1}{2}(c_{CR\_ST\leftrightarrow SY} R_{th\_SY} + R_{th\_ST}) \tag{6.8}$$

In analogy, conductions between stator teeth and outer can $R_{th\_ST\_OC}$, between rotor teeth and inner can $R_{th\_IC\_RT}$, and between rotor teeth and rotor yoke $R_{th\_RT\_RY}$ are calculated, which are

$$R_{th\_ST\_OC} = \frac{1}{2}(c_{CR\_OC\rightarrow ST} R_{th\_OC} + R_{th\_ST}) \tag{6.9}$$

$$R_{th\_IC\_RT} = \frac{1}{2}(c_{CR\_IC\rightarrow RT} R_{th\_IC} + R_{th\_RT}) \tag{6.10}$$

$$R_{th\_RT\_RY} = \frac{1}{2}(c_{CR\_RY\rightarrow RT} R_{th\_RY} + R_{th\_RT}) \tag{6.11}$$

$$c_{CR\_IC\rightarrow RT} + c_{CR\_RY\rightarrow RT} = 1 \tag{6.12}$$

where R_{th_OC}, R_{th_IC}, R_{th_RT} and R_{th_RY} are cylinder based resistances of the regarded component and can be obtained based on (6.5) or (6.6), whereas, $c_{CR\_IC\rightarrow RT}$ and $c_{CR\_RY\rightarrow RT}$ are constant ratios.

6.2.3.2 Convection

The following resistances $R_{th\_fr\_am}$, $R_{th\_OC\_SA}$, $R_{th\_wi\_SA}$ and $R_{th\_OC\_IC}$ fall into this category. The first 3 ones involve air-cooling while the last one may be liquid cooling. For air cooling, convection resistance R_{th_cv} can be modeled by $R_{th\_cv} = 1/(h_{cv}A_{cv})$, where h_{cv} is heat transfer coefficient and A_{cv} is the convection area. Note that "*cv*" in subscript is short for convection. To be simplified, h_{cv} changes approximately proportionally with the small cooling air velocities v up to [15]

$$h_{cv} = 15.5(0.29v + 1) \ \text{in}\,\text{Wm}^{-2}\text{K}^{-1}, \quad v_{\max} = 7.5\text{m/s} \tag{6.13}$$

Resistances between frame and ambient $R_{th\_fr\_am}$ or between the outer can and slot air $R_{th\_OC\_SA}$ is calculated from convection interface R_{h_cv}. As one of the connected components is always air, expressions can be further simplified. For example, the resistance between frame and ambient is written as

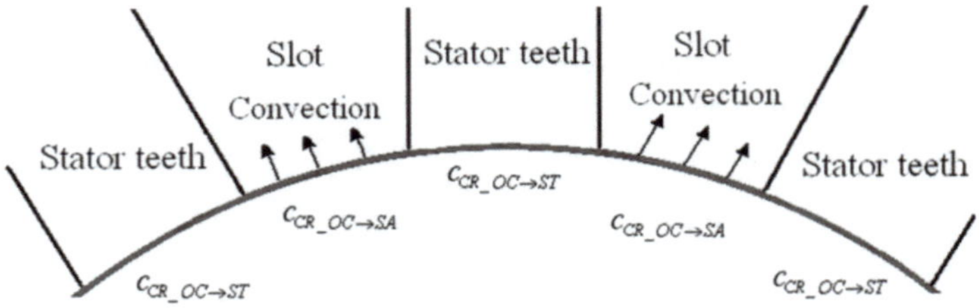

Fig. 6.6 The contact coefficient relationship between the outer can and stator teeth

$$\begin{aligned} R_{th\_fr\_am} &= R_{th\_cv\_fr\_am} + \frac{1}{2}\left(R_{th\_fr} + R_{th\_air}\right) \\ &= \frac{1}{h_{cv\_fr\_am}A_{fr\_am}} + \frac{1}{2}\left(R_{th\_fr} + R_{th\_air}\right) \\ &= \frac{1}{h_{cv\_fr\_am}A_{fr\_am}} + \frac{1}{2}R_{th\_fr} \end{aligned} \quad (6.14)$$

where A_{fr_am} is the contact area between frame and ambient. Similarly, the resistance between outer can and stator slot air is written as

$$\begin{aligned} R_{th\_OC\_SA} &= R_{th\_cv\_OC\_SA} + \frac{1}{2}\left(c_{CR\_OC\to SA}R_{th\_OC} + R_{th\_air}\right) \\ &= \frac{1}{h_{cv\_OC\_SA}A_{OC\_SA}} + \frac{1}{2}\left(c_{CR\_OC\to SA}R_{th\_OC} + R_{th\_air}\right) \\ &= \frac{1}{h_{cv\_OC\_SA}A_{OC\_SA}} + \frac{1}{2}c_{CR\_OC\to SA}R_{th\_OC} \end{aligned} \quad (6.15)$$

where A_{OC_SA} is the convection area between outer can and slot air. The coefficient $c_{CR\_OC\to SA}$ refers to the percentage of contact between outer can and slot air measured from the circumference. The percentage is illustrated in Fig. 6.6, and the following relationship is established as

$$c_{CR\_OC\to SA} + c_{CR\_OC\to ST} = 1 \quad (6.16)$$

Liquid cooling as another form of convection may occur in airgap. Airgap liquid takes away heat from both cans, and is seen turbulent due to rotation. The cylindrical airgap resistance R_{th_AG} is likewise calculated but with k_{eff_liq} added as effective thermal conductivity considering turbulence, the characteristic feature compared with (6.5). For the worst case, the static state is considered and k_{eff_liq} is reduced to the normal material conductivity of liquid k_{liq}. The airgap resistance is then written as (where R_{RT1} is the rotor bore outer radius and δ is the airgap radial length)

$$R_{th\_AG} = \frac{\ln((R_{RT1} + \delta)/R_{RT1})}{2\pi k_{eff\_liq} L_z} \tag{6.17}$$

Now the thermal resistance between cans $R_{th\_OC\_IC}$ can be calculated using half of the resistance from both cans together with the air gap resistance as

$$R_{th\_OC\_IC} = R_{th\_AG} + \frac{1}{2}(R_{th\_OC} + R_{th\_IC}) \tag{6.18}$$

6.2.3.3 Contact

Following resistances $R_{th\_SY\_fr}$, $R_{th\_RY\_sh}$, $R_{th\_wi\_ST}$, $R_{th\_wi\_SY}$ are concerned. The contact resistance R_{th_ct} at the connection interface is from $R_{th\_ct} = 1/(h_{ct} A_{ct})$, where h_{ct} is the contact quality coefficient and A_{ct} is the contact area. Note that subscript "*ct*" is short for contact. Resistances between stator yoke and frame $R_{th\_SY\_fr}$ and between rotor yoke and shaft $R_{th\_RY\_sh}$ fall into this category. Specifically, the contact resistance between stator yoke and frame $R_{th\_SY\_fr}$ and between shaft and ambient $R_{th\_sh\_am}$ are respectively written as

$$\begin{aligned} R_{th\_SY\_fr} &= R_{th\_ct\_SY\_fr} + \frac{R_{th\_SY} + R_{th\_fr}}{2} \\ &= \frac{1}{h_{ct\_SY\_fr} A_{OSSY}} + \frac{R_{th\_SY} + R_{th\_fr}}{2} \end{aligned} \tag{6.19}$$

$$\begin{aligned} R_{th\_RY\_sh} &= R_{th\_ct\_RY\_sh} + \frac{R_{th\_RY} + R_{th\_sh}}{2} \\ &= \frac{1}{h_{ct\_RY\_sh} A_{ISRY}} + \frac{R_{th\_RY} + R_{th\_sh}}{2} \end{aligned} \tag{6.20}$$

where A_{OSSY} and A_{ISRY} stand for contact area of Outer Surface of Stator Yoke and Inner Surface of Rotor Yoke respectively.

6.2.3.4 Compensation Elements

All compensation elements t are listed in Nomenclature. For any couple of connecting components A and B, there are two compensation elements, namely $t_{A \to B}$ and $t_{B \to A}$. Note that the subscript "→" is directional. $t_{A \to B}$ means, from A to B. $t_{B \to A} \neq t_{A \to B}$. The calculation is based on (6.1) and (6.2), e.g. t starting from component A is written as

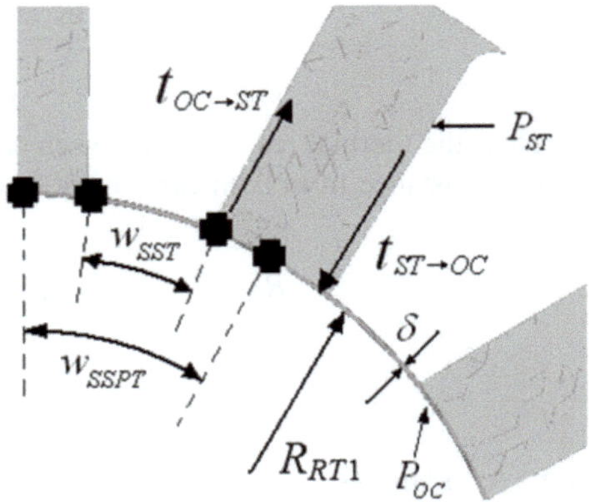

Fig. 6.7 Compensation elements by a couple of connected components with a contact coefficient

$$t_{A\to B} = (P_A/4)(R_{th\_A}/2) \tag{6.21}$$

The compensation is calculated once resistance R_{th} and heat source P are known. Note that there does not exist t from airgap, frame or shaft due to absence of heat sources. For t with stator/rotor teeth (*ST*/*RT*) as terminal, including $t_{IC \to RT}$, $t_{RY \to RT}$, $t_{SY \to ST}$, and $t_{OC \to ST}$, modifications are likewise made due to slots. For example, the contact between stator teeth and outer can is taken in Fig. 6.7, with geometry and heat source shown. All teeth are considered collectively as one component while the can is taken as another. The compensation from stator teeth to outer can is written as

$$t_{ST\to OC} = \frac{P_{ST}}{4}\frac{R_{th\_ST}}{2} \tag{6.22}$$

By contrast, compensation element from outer can to stator teeth $t_{OC \to ST}$ is calculated with some modifications due to the partial contact. For that the contact starting from teeth takes only partial circumference, therefore a coefficient $c_{CR\_OC\to ST}$ is introduced. Based on Fig. 6.6, it is defined as the ratio between the outer can and stator teeth A_{OC_ST} against outer surface of the outer can A_{OSOC}, which is expressed as

$$\begin{aligned} c_{CR\_OC\to ST} &= \frac{A_{OC\_ST}}{A_{OSOC}} = \frac{N_s(w_{SSPT} - w_{SST})L_z}{2\pi(R_{RT1} + \delta)L_z} \\ &= \frac{w_{SSPT} - w_{SST}}{w_{SSPT}} \end{aligned} \tag{6.23}$$

where w_{SSPT} and w_{SST} stand for Stator Slot Pitch at the Top and width of Stator Slot at the Top respectively. $(R_{RT1} + \delta)$ is the outer radius of the outer can, N_s is the number of stator teeth. Further, the compensation element from outer can to stator teeth is then written as

$$t_{OC\to ST} = c_{CR\_OC\to ST}\frac{P_{OC}}{4}\frac{R_{th\_OC}}{2} \tag{6.24}$$

Except the following 6 ones which are related with windings or end windings as

$$t_{EW\to wi}, t_{wi\to ST}, t_{wi\to SY}, t_{wi\to EW}, t_{ST\to wi}, t_{SY\to wi}$$

all the compensation elements in Table 6.2 can be calculated by (6.22) or (6.24), which are listed as follows

$$t_{OC\to AG} = \frac{P_{OC}}{4}\frac{R_{th\_OC}}{2} \tag{6.25}$$

$$t_{IC\to AG} = \frac{P_{IC}}{4}\frac{R_{th\_IC}}{2} \tag{6.26}$$

$$t_{IC\to RT} = c_{CR\_IC\to RT}\frac{P_{IC}}{4}\frac{R_{th\_IC}}{2} \tag{6.27}$$

$$t_{RT\to IC} = \frac{P_{RT}}{4}\frac{R_{th\_RT}}{2} \tag{6.28}$$

$$t_{RT\to RY} = \frac{P_{RT}}{4}\frac{R_{th\_RT}}{2} \tag{6.29}$$

$$t_{RT\to IC} = \frac{P_{RT}}{4}\frac{R_{th\_RT}}{2} \tag{6.30}$$

$$t_{RY\to RT} = c_{CR\_RY\to RT}\frac{P_{RY}}{4}\frac{R_{th\_RY}}{2} \tag{6.31}$$

$$t_{RY\to sh} = \frac{P_{RY}}{4}\frac{R_{th\_RY}}{2} \tag{6.32}$$

$$t_{SY\to ST} = c_{CR\_SY\to ST}\frac{P_{SY}}{4}\frac{R_{th\_SY}}{2} \tag{6.33}$$

$$t_{ST\to SY} = \frac{P_{ST}}{4}\frac{R_{th\_ST}}{2} \tag{6.34}$$

$$t_{SY\to fr} = \frac{P_{SY}}{4}\frac{R_{th\_SY}}{2} \tag{6.35}$$

6.2.4 Model of Coils

6.2.4.1 Overview

The machine operates with large heat dissipation in stator slots, and the main heat flow passes into stator teeth via circumferential direction, leading to characteristic

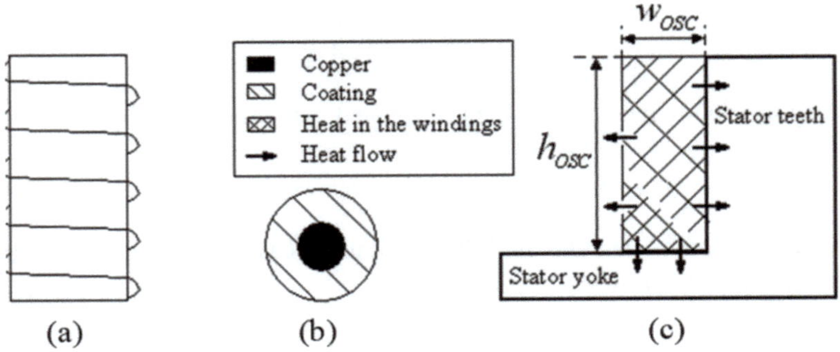

Fig. 6.8 Model of a coil, (**a**) Configuration, (**b**) Cut of a turn, (**c**) Heat flow

features that are discussed in this section. To account for the end effect that both axial sides have different thermal features, coils are further divided into the windings and end windings. Heat transfer concerning windings mainly donates physical contact with lamination while end windings to the cooling at the end part. The winding model is shown in Fig. 6.8a, b respectively. The copper helically wound onto the teeth is coated and thus electrically isolated. The equivalent conductivity depends on factors such as the coating material and integration condition. The heat flow from windings is shown in Fig. 6.8c that besides slot convection, copper loss flows both into teeth and yoke. The portion of each flow is approximated by the contact area as an ideal condition.

The heat flow between windings and stator yoke Q_{wi_SY}, or between windings and stator teeth Q_{wi_ST} can be obtained from copper loss P_{wi} after subtracting the heat removal by slot convection Q_{wi_am}, which are respectively written as

$$Q_{wi\_SY} = a(P_{wi} - Q_{wi\_am}) \tag{6.36}$$

$$Q_{wi\_ST} = (1 - a)(P_{wi} - Q_{wi\_am}) \tag{6.37}$$

$$a = \frac{0.5w_{OSC}^2}{w_{OSC}h_{OSC}} = \frac{w_{OSC}^2}{2w_{OSC}h_{OSC}} = \frac{w_{OSC}}{2h_{OSC}} \tag{6.38}$$

where w_{OSC} is the width of One Side of a Coil (Fig. 6.8c), and the coefficient a from (6.38) refers to the percentage of heat flowing into the yoke. The heat flow between windings and ambient Q_{wi_am} is from Newton's Law as

$$\frac{Q_{wi\_am}}{A_{wi\_am}} = \frac{Q_{wi\_am}}{h_{OSC}L_z} = h_{cv}\Delta T_{wi\_am} \tag{6.39}$$

where ΔT_{wi_am} is the temperature difference between ambient and the surface being cooled, A_{wi_am} is the convection area between windings and ambient, and h_{OSC} is the height of One Side of a Coil.

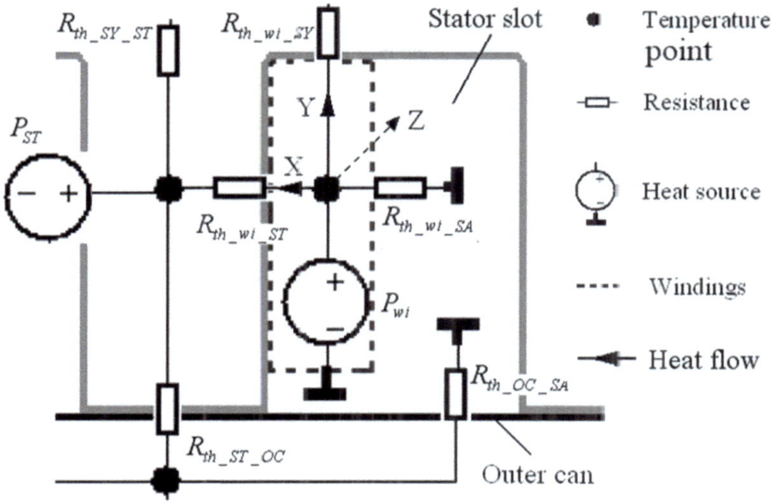

Fig. 6.9 Model of windings with X-Y-Z directions defined

6.2.4.2 Resistances in the Windings

Figure 6.9 shows detailed thermal resistances concerning armature windings. Directions of heat flows are labeled with X-Y-Z Cartesian coordinate located in the nodal of windings. $R_{th\_wi\_ST}$ stands for heat flow into the stator teeth, $R_{th\_wi\_SY}$ for heat flow into the stator yoke, and $R_{th\_wi\_SA}$ for heat removed by slot air. Each is calculated by connected components and contact interfaces as

$$R_{th\_wi\_ST} = R_{th\_ct\_wi\_ST} + \frac{1}{2}\left(R_{th\_X\_wi} + c_{CR\_wi\rightarrow ST}R_{th\_X\_ST}\right) \tag{6.40}$$

$$R_{th\_wi\_SY} = R_{th\_ct\_wi\_SY} + \frac{1}{2}\left(R_{th\_Y\_wi} + c_{CR\_SY\rightarrow wi}R_{th\_Y\_SY}\right) \tag{6.41}$$

$$\begin{aligned} R_{th\_wi\_SA} &= R_{th\_cv\_wi\_SA} + \frac{1}{2}(R_{th\_X\_wi} + R_{th\_air}) \\ &= R_{th\_cv\_wi\_SA} + \frac{1}{2}R_{th\_X\_wi} \end{aligned} \tag{6.42}$$

Subscripts X and Y stand for directions of heat flow of Cartesian coordinate. In the following, each parameter in (6.40), (6.41) and (6.42) is deduced to calculate the connected resistances concerning windings. As the first step, resistance of each component is calculated, including windings, stator teeth and stator yoke. Resistances of windings measured from along each direction are written as

$$R_{th\_X\_wi} = \frac{w_{OSC}/2}{k_{c\_wi}A_{X\_wi}} = \frac{w_{OSC}}{2k_{c\_wi}h_{OSC}L_z} \tag{6.43}$$

$$R_{th\_Y\_wi} = \frac{h_{OSC}/2}{k_{c\_wi}A_{Y\_wi}} = \frac{h_{OSC}}{2k_{c\_wi}w_{OSC}L_z} \tag{6.44}$$

$$R_{th\_Z\_wi} = \frac{L_z/2}{k_{c\_wi}A_{Z\_wi}} = \frac{L_z}{2k_{c\_wi}w_{OSC}h_{OSC}} \tag{6.45}$$

where k_{c_wi} is equivalent thermal conductivity of windings by considering coating and copper as a homogenous body.

Simultaneously, other components that relate to windings are deduced. $R_{th\_X\_ST}$ stands for the resistance measured from X-direction and is used when heat flows from windings to stator teeth. Considering symmetry, it can be written as

$$\begin{aligned} R_{th\_X\_ST} &= \frac{1}{2N_s}\frac{w_{ST}/2}{k_{c\_m}A_{X\_ST}} \\ &= \frac{1}{2N_s}\frac{w_{ST}/2}{k_{c\_m}(R_{ST1} - R_{SY2})L_z} \\ &= \frac{w_{ST}}{4N_s k_{c\_m}(R_{ST1} - R_{SY2})L_z} \end{aligned} \tag{6.46}$$

where (R_{ST1}–R_{SY2}) and w_{ST} are height and width of teeth, $2N_s$ means the number of connected faces. k_{c_m} is the thermal conductivity of the lamination. Similarly, $R_{th\_Y\_SY}$ stands for resistance measured from Y-direction and is used when heat flows from windings to stator yoke, which is

$$R_{th\_Y\_SY} = R_{th\_SY} = \frac{\ln(R_{SY1}/R_{SY2})}{2\pi k_{c\_m}L_z\lambda} \tag{6.47}$$

Then, the contact ratios originating from windings are deduced. $c_{CR\_wi\rightarrow ST}$ stands for the ratio by interface between windings and stator teeth A_{wi_ST}, against the interface between stator teeth and slot A_{ST_sl}. It is written as

$$c_{CR\_wi\rightarrow ST} = \frac{A_{wi\_ST}}{A_{ST\_sl}} = \frac{h_{OSC}L_z}{(R_{ST1} - R_{SY2})L_z} = \frac{h_{OSC}}{R_{ST1} - R_{SY2}} \tag{6.48}$$

In an ideal case, the windings fully contact with teeth, and $c_{CR\_wi\rightarrow ST} = 1$. However in practice, some space at top of the teeth should be considered for cooling when cans are fixed. Similarly, $c_{CR\_SY\rightarrow wi}$ stands for the ratio by interface between windings and stator yoke A_{wi_SY}, against the Inner Surface of Stator Yoke A_{ISSY}. The coefficient depends on radial thickness of the windings and is written as

$$c_{CR\_SY\rightarrow wi} = \frac{A_{wi\_SY}}{A_{ISSY}} = \frac{2N_s w_{OCS}}{2\pi R_{ST1}} = \frac{N_s w_{OCS}}{\pi R_{ST1}} \tag{6.49}$$

Finally, convection or contact resistances $R_{th\_cv\_wi\_SA}$, $R_{th\_ct\_wi\_ST}$ and $R_{th\_ct\_wi\_SY}$ are calculated. The convection between windings and slot air is written as

$$R_{th\_cv\_wi\_SA} = \frac{1}{h_{cv\_wi\_SA}A_{wi\_SA}} = \frac{1}{h_{cv\_wi\_SA}h_{OSC}L_z} \tag{6.50}$$

The contact between windings and stator teeth is written as

$$R_{th\_ct\_wi\_ST} = \frac{1}{2N_s}\frac{w_{OSC}}{h_{ct\_wi\_ST}A_{wi\_ST}} = \frac{w_{OSC}}{2N_s h_{ct\_wi\_ST}h_{OSC}L_z} \tag{6.51}$$

where A_{wi_ST} is contact area between windings and stator teeth, and similarly $2N_s$ means number of contact surfaces connected in parallel. Equation (6.21) is simple but requires determination of $h_{ct\_wi\_ST}$, the contact coefficient between windings and teeth. Similarly, contact resistance between windings and stator yoke is written as

$$R_{th\_ct\_wi\_SY} = \frac{1}{2N_s}\frac{h_{OSC}}{h_{ct\_wi\_SY}A_{wi\_SY}} = \frac{h_{OSC}}{2N_s h_{ct\_wi\_SY}w_{OSC}L_z} \tag{6.52}$$

Likewise, the coefficient $h_{ct\_wi\_SY}$ is determined by experiment. Usually, $h_{ct\_wi\_SY}$ is numerically much lower than $h_{ct\_wi\_ST}$ because the windings are not tightly connected onto stator yoke.

6.2.4.3 Resistances in the End Windings

Apart from air cooling, some heat flow passes from the end windings through windings into stator and further into frame. The thermal resistance from the midpoint of end windings to windings is deduced. Using symmetry, there are 4 connections in parallel in one coil. For totally N_s coils the resistance $R_{th\_wi\_EW}$ is written as

$$\begin{aligned} R_{th\_wi\_EW} &= \frac{1}{4N_s}\frac{1}{2}(R_{th\_Z\_wi} + R_{th\_EW}) \\ &= \frac{1}{8N_s}\left(\frac{L_z/2}{k_{c\_wi}w_{OSC}h_{OSC}} + \frac{L_{EW}/2}{k_{c\_wi}w_{OSC}h_{OSC}}\right) \\ &= \frac{L_z + L_{EW}}{16N_s k_{c\_wi}w_{OSC}h_{OSC}} \end{aligned} \tag{6.53}$$

where L_{EW} is average length of the end windings, A_{wi_EW} is the conduction area between windings and end windings and is written as

$$A_{wi\_EW} = w_{OSC} h_{OSC} \tag{6.54}$$

6.2.4.4 Compensation Elements of a Coil

Due high amount of copper loss as heat source, compensation elements regarding a coil start from that coil towards a neighboring component in one direction. Therefore according to the calculation principle, resistance and heat source of a coil are used to get the compensation values. The main modification is that the proportion of heat flow a whether into windings or end windings should be predetermined. Resistances between the windings and connected components are now written as

$$t_{wi \to ST} = \frac{Q_{wi\_ST}}{4} \frac{R_{th\_X\_ST}}{2} = \frac{(1-a)(P_{wi} - Q_{wi\_Am})}{4} \frac{R_{th\_X\_ST}}{2} \tag{6.55}$$

$$t_{wi \to SY} = \frac{Q_{wi\_SY}}{4} \frac{R_{th\_Y\_wi}}{2} = \frac{a(P_{wi} - Q_{wi\_Am})}{4} \frac{R_{th\_Y\_wi}}{2} \tag{6.56}$$

$$t_{wi \to EW} = \frac{(P_{wi} - Q_{wi\_Am})}{4} \frac{R_{th\_Z\_wi}}{2} \tag{6.57}$$

Heat flow between windings and ambient Q_{wi_am} is calculated by (6.39), based on the temperature difference between windings and the cooling medium. Therefore, the winding temperature is one of the parameters required to be simulated as prerequisite. To solve this problem, a tentative initial value is used, and Q_{wi_am} value is iteratively modified based on simulation results from the thermal network.

The rest compensation elements concerning windings are written as

$$t_{EW \to wi} = \frac{(P_{EW} - Q_{EW\_am})}{4} \frac{R_{th\_EW}}{2} \tag{6.58}$$

$$t_{SY \to wi} = c_{CR\_SY \to wi} \frac{P_{SY}}{4} \frac{R_{th\_SY}}{2} \tag{6.59}$$

Q_{EW_am} stands for the amount of heat removed through convection in the end windings. Similarly, it is obtained when the cooling condition h_{cv} and temperature difference values are known. The machine end part is further discussed in Sect. 6.2.5. As to compensation from stator teeth to windings $t_{ST \to wi}$, almost heat from stator teeth P_{ST} goes directly into yoke, instead of into the windings. Therefore, the heat flow from stator teeth to windings $P_{ST \to wi}$ is negligible. The compensation is now written as

$$t_{ST \to wi} = \frac{P_{ST \to wi}}{4} \frac{R_{th\_ST}}{2} = 0 \tag{6.60}$$

In fact P_{ST} has been already used to calculate $t_{ST \to SY}$ and $t_{ST \to OC}$.

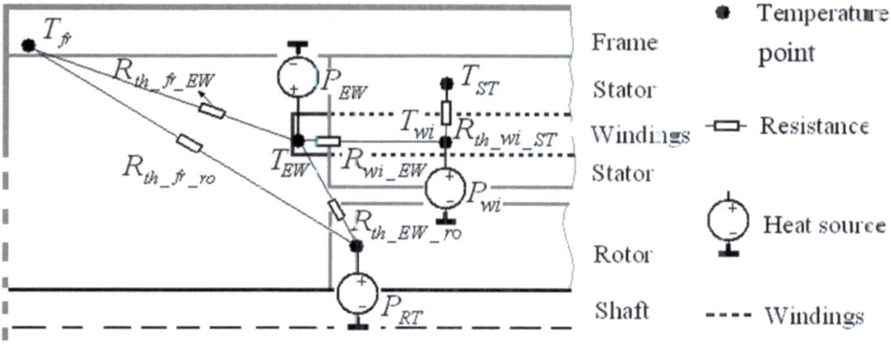

Fig. 6.10 Heat flow in the machine end part

Table 6.5 Thermal resistances in the machine end part

Symbols	Resistance between	Type
$R_{th\_fr\_EW}$	Frame and end windings	Convection
$R_{th\_fr\_ro}$	Frame and rotor	Convection
$R_{th\_EW\_ro}$	End windings and rotor	Convection

6.2.4.5 Model of the End Part

Heat flow in the machine end part is usually subjected to forced convection, quite a complicated process especially when combined with turbulent flow and complex geometry. Generally, faces of the rotor at the end part and the end windings are mutually affected and the heat goes to ambient through frame. In Fig. 6.10, heat flow relation is simplified by a triangle network. The resistances concerning forced convection between frame, rotor and end windings are defined in Table 6.5. Such parameters are determined only by experiment.

6.3 Application and Discussion

Before the calculation, all material properties are defined in Table 6.6, in which abbreviations in subscript stand for corresponding component names. Conductivity value of armature coils is considered 1/10 of copper. The cooling fluid between cans is set as static water, the worst case in which the minimum loss is taken away. Loss values from regarded components as heat source are shown in Table 6.7, in which the can losses are remarkably high. The use of cans and consequent can loss characterizes temperature rise and heat transfer. Calculation results of all elements are shown in Tables 6.8 and 6.9.

Ambient temperature is set 0 °C to calculate just temperature rise. The heat pathway is shown in Fig. 6.11, in which only heat sources and resistances are presented. The can loss is partly taken away by cooling $R_{th\_OC\_SA}$ whereas the rest

Table 6.6 List of thermal conductivities

Thermal conductivity k_c	Values ($W \cdot m^{-1} \cdot K^{-1}$)
k_{c_m}	54
k_{c_cop}	400
k_{c_wi}	40
k_{c_can}	16.3
k_{c_air}	0.03
k_{c_liq}	0.5
$k_{c\_eff\_liq}$	0.5

Table 6.7 List of heat sources (time averaged values)

Heat source P	Calculated values (W)
P_{SY}	66.5
P_{ST}	71.1
P_{wi}	541.4
P_{EW}	50.6
P_{OC}	1120
P_{IC}	1118
P_{RT}	20.2
P_{RY}	17.1

Table 6.8 Calculation results of all resistances in the thermal network

Thermal resistance R_{th}	Calculated values (°C/W)
$R_{th\_fr\_am}$	0.0017
$R_{th\_SY\_fr}$	0.00459
$R_{th\_SY\_ST}$	0.017
$R_{th\_wi\_ST}$	0.002
$R_{th\_wi\_SY}$	0.0055
$R_{th\_wi\_EW}$	0.0104
$R_{th\_ST\_OC}$	0.017
$R_{th\_OC\_IC}$	0.022
$R_{th\_IC\_RT}$	0.0034
$R_{th\_RT\_RY}$	0.0071
$R_{th\_RY\_sh}$	8.0
$R_{th\_wi\_SA}$	1.256
$R_{th\_OC\_SA}$	21.85

passes into stator teeth, and further yoke and frame. Copper loss, apart from slot cooling, in analogy goes into stator teeth and yoke, and the nodal temperature of armature coils highly depends on contact interface with lamination.

The network consisting resistances R_{th}, heat sources P (voltage generator), compensation elements t (voltage potential) and heat flow (current direction) is shown in Fig. 6.12. Voltmeters as a means of measurement are used. At each dot, nodal temperature of each component T_i is recorded. The machine end part convection is marked as well.

Table 6.9 Calculation results of all compensation elements in the thermal network

Compensation element t	Calculated values (°C)
$t_{ST \to OC}$	0.347
$t_{OC \to ST}$	0.0368
$t_{OC \to AG}$	0.11
$t_{IC \to AG}$	0.11
$t_{IC \to RT}$	0.0368
$t_{RT \to IC}$	0.0159
$t_{RT \to RY}$	0.0159
$t_{RY \to RT}$	0.0157
$t_{RY \to sh}$	0.047
$t_{EW \to wi}$	0.263
$t_{wi \to EW}$	0.667
$t_{wi \to ST}$	2.336
$t_{ST \to wi}$	0
$t_{SY \to ST}$	0.0199
$t_{ST \to SY}$	0.347
$t_{SY \to wi}$	0.347
$t_{wi \to SY}$	0.023
$t_{SY \to fr}$	0.0457

The FE method is first applied as a means of verification, with identical heat transfer and physical cooling condition applied for both methods. In FE, eddy current loss density is seen uniformly distributed within regarded components. The proposed network does not include temperature variation in axial direction. For this reason, 2D FE is used to compare. However, heat flow from end windings is not considered in FE. To solve such inconsistency, copper loss is pre-determined. First, heat flow through the machine end part is recorded by the proposed thermal network. Then total copper loss subtracting power losses that flow throughout the end regions is applied as load in the FE. The FE is under the condition: frame cooling $h_{cv\_fr\_am} = 12$ k(Wm^{-2} K^{-1}), natural slot convection $h_{cv\_wi\_SA} = h_{cv\_OC\_SA} = 15.5$ (Wm^{-2} K^{-1}) and static coolant in airgap $k_{eff\_liq} = 0.5$(Wm^{-1} K^{-1}) as the worst case. The temperature distribution inside the machine is shown in Fig. 6.13. Generally, temperature value decreases from rotor toward ambient due to frame convection. There exists a large numerical gap at the boundary between the stationary and rotary components due to the considerable loss from cans. In other words, temperature rise especially the rotor part is greatly enhanced by the use of cans. In particular, the most thermal sensitive region of a can locates on mid of a rotor slot. This is because such region lacks direct contact with poles and hence hinters heat transfer.

A detailed heat transfer is shown in Fig. 6.14. At the overlap zone of stator and rotor teeth, the generated can loss as source of heat flow directly goes into stator teeth. However at the zone where cans and slot air are contacted, the generated heat flow from the can loss first chooses the circumference along the can shield and then goes forward into stator teeth, forming the nearest pathway to ambient. Due to high temperature level as well as a narrow pathway of the can shield, the value of flow

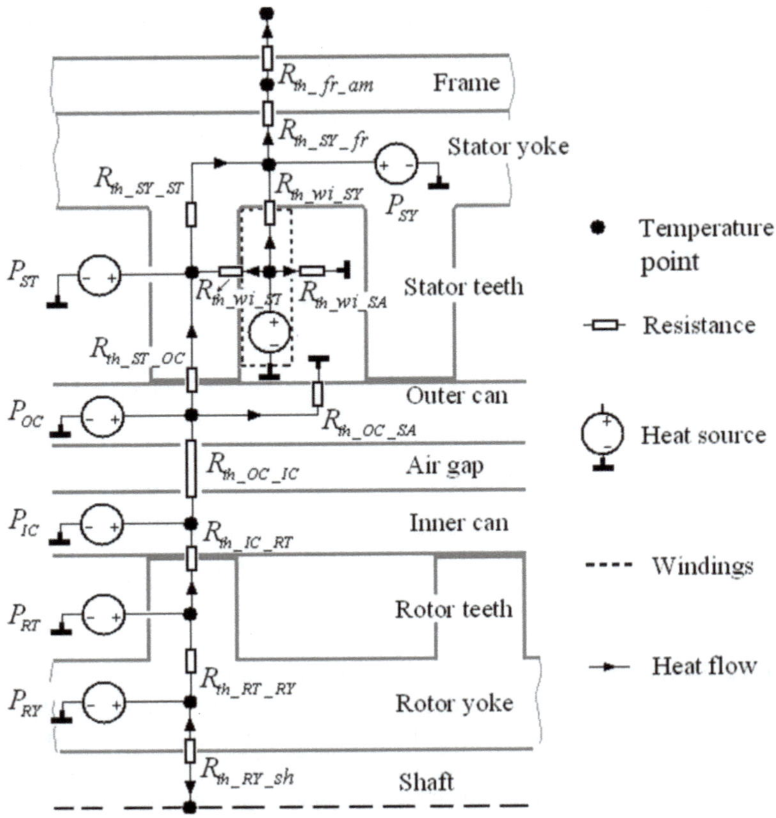

Fig. 6.11 Detailed heat flow of the proposed network model

intensity is remarkably higher, which may cause local overheating. Meanwhile, a small portion of heat flow chooses pathway of rotor teeth back and then forward into stator. This is because of the high heat flow intensity on cans that affect the paths.

Nodal temperature value of each component is calculated respectively by the conventional and improved network, and FE is used as reference. For FE, the temperature value is averaged based on all discretized elements. A numerical agreement between FE and improved network is shown in Table 6.10, justifying the necessity of using compensation elements. The temperature discrepancy of the traditional model is obvious in windings, because the calculated compensation element from windings $t_{wi} \rightarrow {}_{ST}$ in Table 6.9 is comparatively higher, as both P and R values of windings (wi) are remarkable. Therefore, temperature value of windings is more compensated, leading to calculation difference compared with the conventional model. Such numerical discrepancy occurs throughout the entire operation range, particularly when the machine is of high power with P_{wi}.

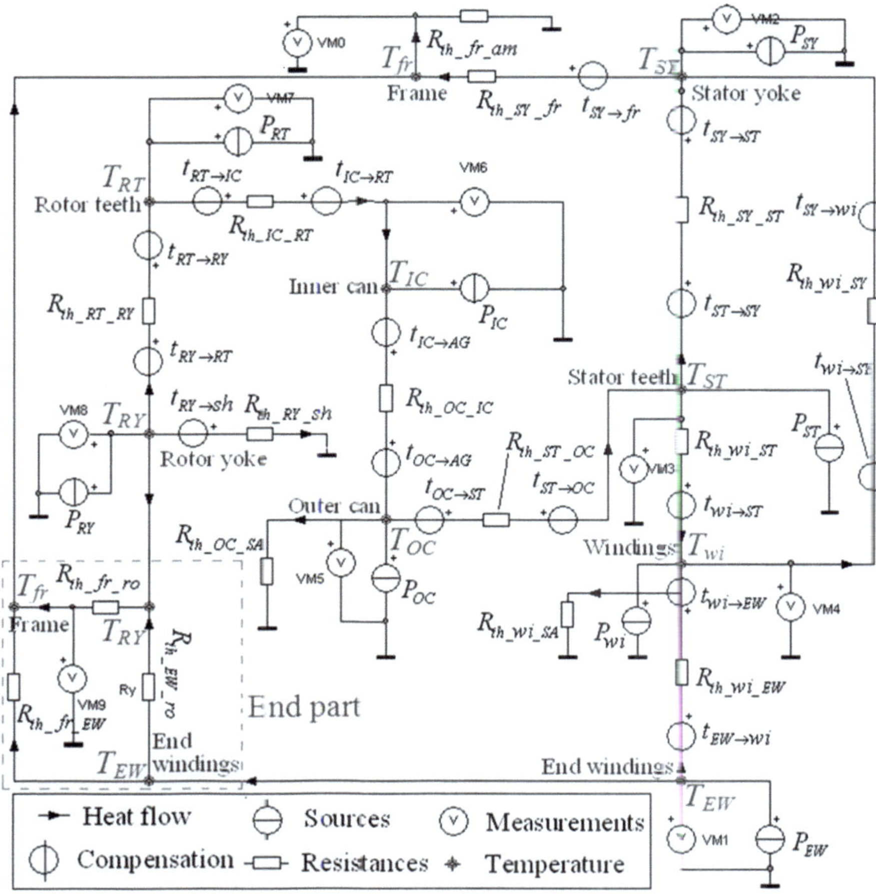

Fig. 6.12 The network overview

For an ordinary machine without cans, coils especially the end part are the most thermal sensitive region with highest temperature rise. For a canned machine, cans are the most critical components due to high can loss and likewise, the end part is critical. However, this part does not set out as a particular concern. To solve this problem, the model is further refined by axially separating the can shield into the imaginary mid, partial and end sectors that are connected. Each sector is imposed with can loss as heat source, which is determined by experience.

To find out thermal distribution on the cans and further verify the proposed model, measurement is taken. Prototype of the studied machine is shown in Fig. 6.15. In Fig. 6.15a the stator can shield is clamped onto the stator bore so that saliency cannot be seen. In Fig. 6.15b the stator before the can is fixed onto the stator bore is shown. In measurement, the machine is running at low speed 500 rpm with DC phase current up to 170A, the equivalent current density value of 16.1A/mm^2.

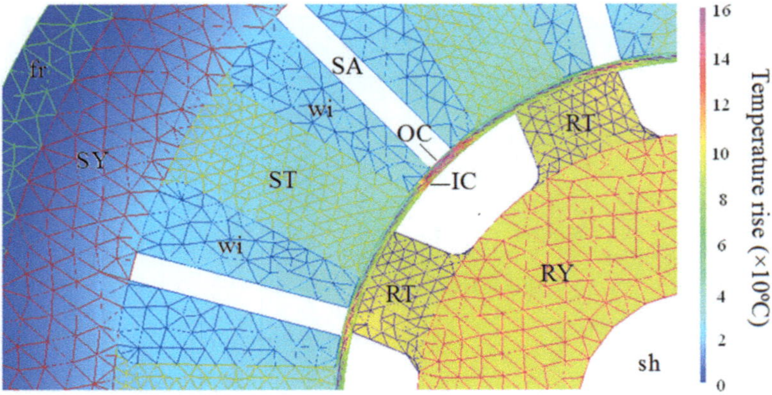

Fig. 6.13 Temperature rise of each component due to the use of cans

Fig. 6.14 Detailed heat flow of the can shields and airgap

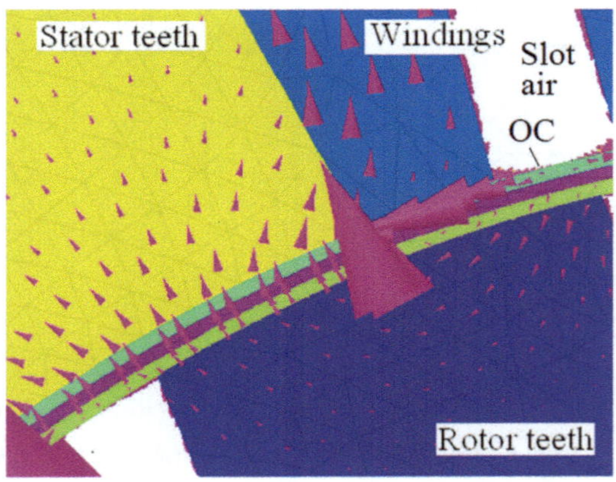

Table 6.10 Component temperature rise by different methods at steady state, with ambient temperature 23.4 °C

	$I_{ph} = 8.1\text{A/mm}^2$			$I_{ph} = 16.1\text{A/mm}^2$		
T (°C)	I	II	III	I	II	III
T_{fr}	26.5	26.5	26.5	29.2	29.2	29.3
T_{EW}	34.9	33.4	33.7	46.2	41.3	41.6
T_{OC_M}	40.4	40.1	39.8	59.9	58.8	58.3
T_{OC_P}	40.6	40.1	39.8	61.0	59.5	58.7
T_{OC_E}	48.0	44.7	43.4	74.1	67.8	66.5

I = Traditional network, II = Proposed network with compensation elements, III = Measurement

Temperature points at typical locations along axial direction of the stator can (outer can) 1–3 are recorded, which are, (1) mid-point T_{OC_M}, (2) partial point T_{OC_P}, (3) end-point T_{OC_E}, and additional (4) end windings T_{EW}.

For sake of comparison, thermal conduction, contact, loss and cooling conditions as input into the network model are identified from test bench. To identify loss of

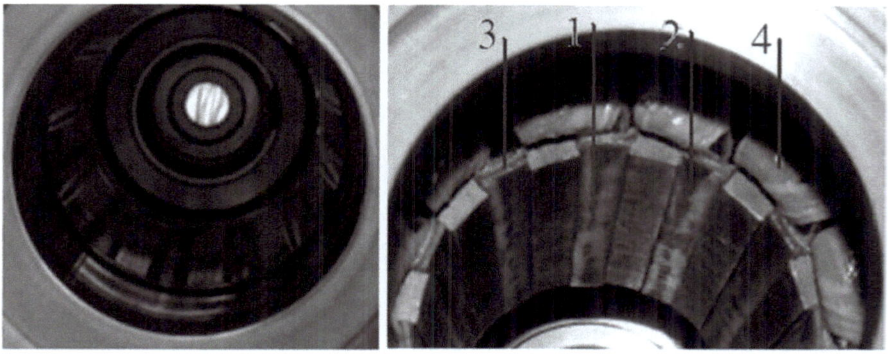

(a) (b)

Fig. 6.15 Stator prototype of the canned machine, (**a**) canned, (**b**) before the stator can is fixed onto the bore, showing locates of thermal sensors, *1*: axially mid, *2*: axially partial, *3*: axially end, *4*: end windings. Note that plastic partitions that help to consolidate armature coils is removed before the stator can is fixed and thermal sensors will be attached on 1–3 of the can

each component, measurement is recorded. Loss includes can loss, iron loss and copper loss. Friction loss is not taken at low speed operation. The total loss is first obtained by the energy conversion principle. Copper loss is then separated out by I^2R when temperature rise is stabilized, remaining total eddy current loss, including can loss and iron loss. To separate iron loss out, another SRM having the same geometry but without cans is built up and iron loss is obtained in analogy by subtracting copper loss from the total loss measured. Hence, the can loss is obtained by subtracting iron loss out from the total eddy current loss.

Temperature rises by both simulation and measurement are compared in Table 6.10, where the stator frame temperature is recorded as potential value. For network simulation, whether compensation elements are used is respectively shown. As to the end windings, it is again shown the use of compensation elements that helps to reduce computation error. Such assumption also holds true on the outer can mid-point T_{OC_M}, partial point T_{OC_P}, and in particular end point T_{OC_E}. The axial end part of the can is the most thermal sensitive region. Measurement shows that the temperature rise at the ending is particularly high and increases sharply with phase current density. This is due to the eddy current swirls on the can concentrates at the endings, causing higher loss density.

Transient analysis is performed by updating the network model with thermal capacitances. The capacitance of each regarded component i is derived from geometrical and material data of the machine, which is

$$C_{th\_i} = m_i c_i \tag{6.61}$$

where m_i is the mass and c_i is the specified heat capacitance of i-th component. Temperature variation of the end part of coils and cans T_{EW} and T_{OC_E} as typical locations are comparatively shown in Fig. 6.16. Use of compensation elements

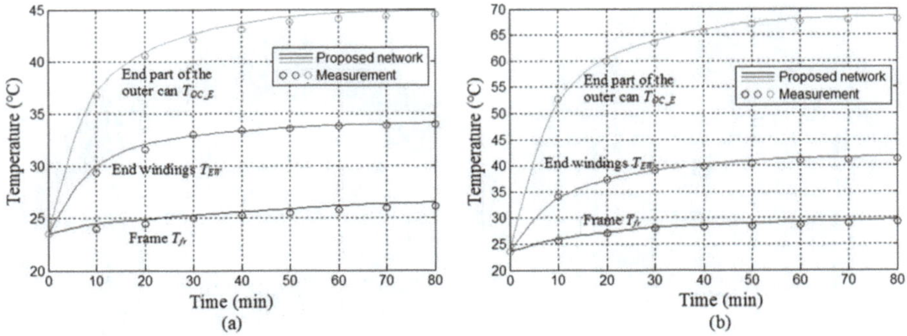

Fig. 6.16 Nodal temperature rise of critical points, (**a**) under phase current $I_{ph} = 8.1$ A/mm^2, (**b**) under $I_{ph} = 16.1$ A/mm^2

Table 6.11 Thermal dependent parameters of the can material

T (°C)	k_c (W/mK)	ρ (μΩm)
0	9.6554	1.2374
100	11.1444	1.8715
200	12.9987	2.5025
300	14.8964	3.1625
400	16.7106	3.8701

improves calculation accuracy by comparing with measurement. The discrepancy in value is acceptable when phase current density I_{ph} varies.

6.4 Electro-thermal Coupled Analysis

6.4.1 The Calculation Flow

The can loss is an undesired feature, resulting in a big temperature jump. In turn, the temperature rise will affect properties of each component, in particular the electrical resistance and thermal conductivity of cans, leading to the change of can loss generated. This would be in summary a thermal and electromagnetic coupling phenomenon that greatly determines airgap flux and thus loss and efficiency of the machine system, which is studied in this section.

In the following simulation, thermal conductivity k_c and electrical resistivity ρ of cans, which are subjected to variation due to a high nodal temperature jump, are assumed temperature dependent. Typical values are shown in Table 6.11. Generally both values rise up linearly with temperature increase, which facilitates convergence of iterative calculation to be illustrated below. The electrical resistivity is more sensitive to temperature, and the value is almost doubled from 0 °C to 200 °C, resulting in a fundamental change of can loss to be illustrated below.

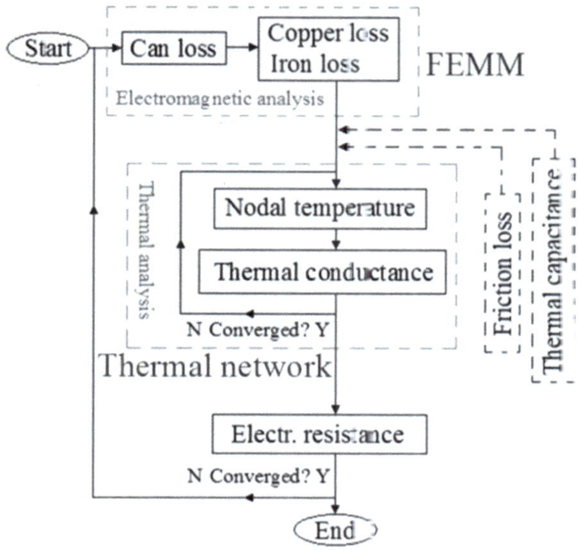

Fig. 6.17 The calculation flow of the coupled analysis

Figure 6.17 shows calculation flow of the electromagnetic thermal coupled analysis. To begin with, thermal conductivity and electrical resistivity at room temperature are taken and electromagnetic analysis as the first stage is carried out. Analysis is performed by FEMM (Finite Element Method Magnetics). This program discretizes the machine into very limited number of meshes, resulting in small computation. The discretization approach is the most reliable to account for the locally distributed magnetic saturation especially on prominent saliencies, which will guarantee accurate estimation of flux linkage that the can loss generation heavily depends on. In the second stage, thermal analysis is performed. Loss calculated for each component is averaged as heat source into the thermal network. First, nodal temperature of each component is calculated based on initial settings of room temperature condition. Then, thermal conductivity value of the cans is iteratively updated. It is found that the temperature rises of both cans of SRMs are similar, as cans are close to each other and work in quite similar electromagnetic environment. Hence, material parameters of both cans are iterated simultaneously to reduce algorithm complexity. When thermal calculation is done, electrical resistivity is updated according to the obtained temperature rise. The initial value is likewise set according to room temperature and if it is not converged with the updated value, the program returns again to the electromagnetic analysis.

The calculation flow is controlled by Matlab that both modules act as sub-models commanded by m-file. In particular, FEMM connects FE with m-file commands conveniently, facilitating flexible logic control. To reduce computation, the maximum number of iteration steps to update thermal resistance is manually limited. In addition, such a calculation flow can be updated to facilitate comparison with measurement by adding mechanical friction loss between can shields, which is

Table 6.12 Steady state can loss (both cans) and temperature variation of the stator can (on average) at different operation condition, with constant speed of 3 krpm

	Loss (W)		T (°C)		
I_{ph}(A)	S	M	S	M	M_e
25	391.5	452.5	33.1	36.1	37.3
60	941.2	1050.0	55.0	60.3	67.3
90	1454.5	1612.6	75.8	79.8	92.9
120	1926.1	2140.1	94.7	101.7	120.0
148	2331.7	2606.5	111.4	118.4	140.5
175	2647.4	2996.0	122.0	130.2	155.4[a]

S simulation, *M* measurement, M_e measurement at the end part of the can
[a]Estimated

determined by rotor inertia, shaft speed and retarding time. Friction loss is characteristic due to closely located cans that have high relative movement.

6.4.2 Application and Discussion

The machine in Fig. 6.2 is modeled and tested. Table 6.12 shows the can loss values as a function of current at constant rotor speed 3krpm. As to simulation, the can material properties are obtained according to recorded temperature values of the test bench at steady state. Due to the temperature rise that in turn affects electrical resistance and thermal conductance, simulation results make some features. First, the loss value is lower at each phase current. Second, loss rises up almost linearly with increase of current, other than geometrically. Both aspects are from the change of the can material property that helps to resist the increase of the induced can loss. Compared with measurement, there exists a difference with regard to loss and the numerical gap is enlarged with higher current. This is mainly because in experiment the eddy current concentrates at the end part of the can, causing more loss.

In terms of temperature rise, it is shown in Table 6.12 that both simulation and measurement values on the stator can shield, a representative component, are similar. Due to additional can loss from the machine end part, the measurement values are higher at each operation condition compared with simulation. In addition, temperature rise at the axial endings of the stator can is recorded. It shows that the end part is the most thermal sensitive that requires scrutiny in operation. It is shown that temperature rise at the end part is higher and the numerical gap is enlarged with higher phase current I_{ph}. This is because when airgap flux is strengthened, the induced eddy current on the can is increased as well. As a result, eddy current at the end part is more concentrated, making a higher temperature rise.

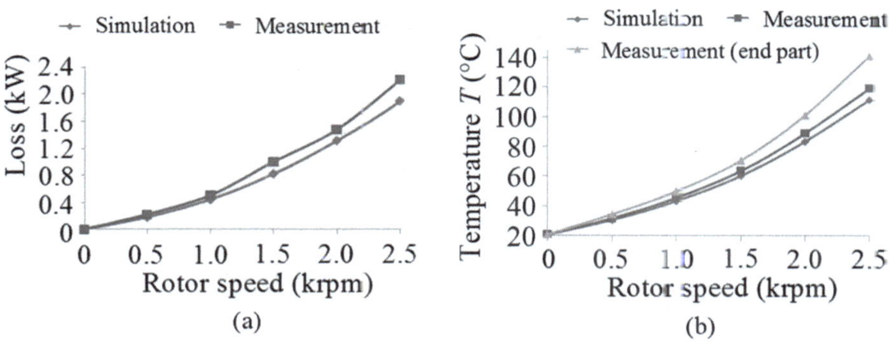

Fig. 6.18 Can loss (**a**) and temperature rise (**b**) with rotation speed 0–2.5 krpm at steady state, the fixed turn-on/off control with 170A hysteresis current level is applied

Loss variation is further illustrated with increasing rotor speed at constant high phase current $I_{ph} = 170\text{A}$. Figure 6.18a shows the can loss value (both cans) geometrically rises up. The numerical gap between simulation and measurement increases with speed, as more loss occurs at the end part. Accordingly temperature rise of the stator can shield is shown in Fig. 6.18b. The rise increases geometrically as well and the measurement is higher due to the end part. It is shown again that the end part is the most thermally critical location.

6.5 Chapter Summary

This chapter investigates temperature rise characteristics of the can shield by exemplifying a novel canned switched reluctance machine. Thermal characteristics are that the temperature rise of each component is raised up especially the rotor part due to considerable loss from the cans, making a substantial difference. In particular axial end part of the can shield is the most thermal sensitive region due to end effect.

The use of compensation elements in the lumped thermal network will improve calculation accuracy. Such a model requires small amount of computation and is transferrable to electrical systems and electromagnetic devices.

The use of cans makes totally different output characteristics. Further as an application, the developed thermal model is coupled with electromagnetic analysis to calculate thermal-electro coupling effect on cans. It is shown that temperature rise will change the material properties, which helps to resist eddy current generation on the can that in turn affect the rise. The calculation flow proposed is time saving and accuracy is verified by comparing with experiments.

References

1. D. Staton, A. Boglietti, A. Cavagnino, Solving more difficult aspects of electric motor thermal analysis in small and medium size industrial induction motors. IEEE Trans. Energy Convers. **20** (3), 620–628 (2005)
2. A. Rasekh, P. Sergeant, J. Vierendeels, Convective heat transfer prediction in disk-type electrical machines. J. Appl. Therm. Eng. **91**(5), 778–790 (2015)
3. K.H. Lee, H.R. Cha, Y.B. Kim, Development of an interior permanent magnet motor through rotor cooling for electric vehicles. J. Appl. Therm. Eng. **96**(25), 348–356 (2016)
4. Y. Huai, R.V.N. Melnik, P.B. Thogersen, Computational analysis of temperature rise phenomena in electric induction motors. J. Appl. Therm. Eng. **23**(7), 779–795 (2003)
5. C. Ferrandi, F. Iorizzo, M. Mameli, S. Zinna, M. Marengo, Lumped parameter model of sintered heat pipe: transient numerical analysis and validation. J. Appl. Therm. Eng. **50**(1), 138–146 (2013)
6. S. Lim, S. Park, H. Chung, M. Kim, Y.J. Baik, S. Shin, Dynamic modeling of building heat network system using Simulink. J. Appl. Therm. Eng. **84**(5), 1280–1290 (2015)
7. Y.L. Low, B.L. Ibrahim, C.P.H. Tan, J.Q. Mou, E.H. Ong, *Thermal Network Model for Temperature Prediction in Hard Disk Drive* (Springer, .Berlin). ISSN:1432-1858. https://doi.org/10.1007/s00542-009-0866-6
8. A. Boglietti, A. Cavagnino, M. Lazzari, A simplified thermal model for variable-speed self-cooled industrial induction motor, International Conference on Electrical Machines, ICEM 2004, Greece
9. C. Gramsch, A. Blaszczyk, H. Lobl, S. Grossmann, *Thermal Network Method in the Design of Power Equipment* (Springer .Berling/Heidelberg). ISBN:978-3-540-71980-9. https://doi.org/10.1007/978-3-540-71980-9_22
10. K. Li, S.P. Wang, J.P. Sullivan, A novel thermal network for maximum temperature-rise of hollow cylinder. J. Appl. Therm. Eng. **52**(1), 198–208 (2013)
11. A.B. Solomon, M. Sekar, S.H. Yang, Analytical expression for thermal conductivity of heat pipe. J. Appl. Therm. Eng. **100**(5), 462–467 (2016)
12. X.H. Hao, B. Peng, Y. Chen, G.N. Xie, Transient thermal model of a permanent magnet synchronous planar motor considering spreading thermal resistance. J. Appl. Therm. Eng. **81** (25), 1–9 (2015)
13. G. Dajaku, D. Gerling, Novel lumped parameter thermal model for electrical systems, European Conference on Power Electronics and Applications, EPE 2005, Germany
14. Q. Yu, C. Laudensack, D. Gerling, Improved lumped parameter thermal model and sensitivity analysis for SR drives, International Conference on Electrical Machines, ICEM 2010, Italy
15. T.J.E. Miller, *Switched Reluctance Motor and Their Control* (Magna Physics Publishing and Clarendon Press, Oxford, 1993). ISBN:0198593872

Chapter 7
Conclusions and Future Work

7.1 Conclusions

This book presents modeling and analysis of electromagnetic and thermal features of canned electrical machines as a hydraulic pump drive. Such a drive integrates the pump body and is capable of working beneath the liquid being pumped with little maintenance. The canned machine is a featured category in the electrical machine drive family that metallic non-laminated can shields are inserted in the magnetically sensitive airgap region. Therefore, electromagnetic and consequent thermal features are of fundamental difference compared with an ordinary one.

The first chapter describes briefly features canned machines and their corresponding applications. Based on the use of cans, recent development and state-of-the-art prototypes are introduced. Advantages and disadvantages of AC or DC excited canned machines are discussed. In particular, salient switched reluctance machine (SRM) as a hydraulic pump drive is shown with simple structure, low cost and high robustness.

The second chapter studies the saliency effect in terms of the electromagnetic analysis. Such effect determines flux linkage characteristics by manipulating equivalent airgap paths that is deemed rotor position dependent. Further when high power density is concerned, local magnetic saturation will complicate the determination of magnetic reluctance. Calculation of the flux linkage and magnetic reluctance is the prerequisite to analyze the can effect and models are introduced based on the classical magnetic equivalent circuit (MEC). Specifically, the calculation accuracy is improved by discretizing the sensitive airgap region and high locally saturated poles into limited number of reluctance elements according to flux path feautres. Such discretization approach accounts for saliency effect with small computation. Alternatively, the discretization is replaced by using curve fitting method, and the model is updated with magnetic phase coupling considered. Further, the use of cans is easily updated into the network model.

Q. Yu et al., *Analysis and Mathematical Models of Canned Electrical Machine Drives*, https://doi.org/10.1007/978-981-13-2745-2_7

The third chapter studies the can effect of a novel salient canned switched reluctance machine. When cans are imposed in alternative electromagnetic fields, eddy current is induced that in turn affects airgap flux distribution. That is, inserting a can in airgap will weaken airgap flux and harmonics, and reduce output torque. The use of cans brings considerable ohmic can loss, and the loss variation is featured by a couple of peaks. The principal part of loss occurs when each phase is in sequence under the progress of phase defluxing while the secondary occurs under phase fluxing. At different control modes, the couple of peaks persists, and average loss values rise up with increase of rotor speed geometrically and with increase of current excitation in a saturated way.

The fourth chapter studies spatial distribution of magneto motive force (MMF), the prerequisite of analytical modeling of the can effect and canned machines. The armature coils are mainly studied, followed by permanent magnets. For armature coils, both tooth concentrated and slot distributed topologies are shown. In particular for a concentrated coil, the inside wire layout is modeled. The model starts from a single turn/coil using Fourier method, and then is extended to all coils in one phase and finally to all phases. In addition, MMF in axial direction is studied using Fourier discretization. The proposed method can have a fast and accurate magnetic analysis and a concrete view on electromagnetic modeling of electrical machines.

The fifth chapter proposes an analytical model of concentric layer structure in 3 dimensions for canned electrical machines. This model divides the machine into airgap, cans, rotor pole, rotor yoke and shaft that are concentric connected cylindrical components. Based on MMF distribution, this model stems from Maxwell Equations in vector form. By separation of variables, magnetic vector potential of each layer is deduced. Loss of each layer including the can loss is calculated by Poynting's theorem of energy conversion. Mechanical torque of each layer is calculated by applying Maxwell stress tensor. This model is convenient to analyze the can effect and at the same time compare with the ordinary machine by simply adding layers in airgap as can shields. Based on this model, the can effect of induction machine, or recently developed permanent magnet machine and switched reluctance machine is shown comparatively.

The sixth chapter investigates the thermal and heat transfer of canned machines, exemplifying a canned switched reluctance machine. It is shown that the considerable ohmic can loss will lead to featured temperature distribution and a numerical jump especially on the rotary components. In particular the axial end part of the can is the most thermal sensitive region. Further as an application, the thermal-electro coupling effect on cans is studied by combining the thermal and electromagnetic analytical models. It is shown that the temperature rise will change the material properties of cans, which helps to resist eddy current generation on the can. Meanwhile in the calculation of nodal temperature rise, the calculation accuracy is improved by adding compensation elements, based on the conventional network. In the past, the systematic mistake is often neglected or rounded using a curve fitting procedure. Such improvement requires small computation and is easily transferrable to electrical systems and electromagnetic devices.

7.2 Future Work

Accurate electromagnetic and thermal analysis of electric machine drives is of great importance for a successful design and effective utilization. Due to the characteristic use of the non-laminated metallic can shields in airgap, both electromagnetic and thermal features of a canned machine are of fundamental difference. Therefore, the aim of this research work is to develop new and effective mathematical models which describe complete performances of the canned electrical machines with high accuracy.

The state-of-the-art machine is expected of high power density and efficiency. In future work, combination of this couple of features will be further studied on canned machines. For one thing, the increase of power density will lead to the sharp jump of can loss, the main constraint for efficiency enhancement. For another, control methods targeting on loss minimization may lead to the reduced power density. By utilizing a novel machine structure as well as flux harmonic reduced control, there would be a compromise between power density and efficiency.

Meanwhile due to harsh environment that the machine is working, little maintenance is required and therefore stable operation and duration are another concern. Due to the very limited airgap space that accommodates both cans of extremely thin shields, together with cooling liquid going through, slight manufacturing inaccuracy, assembly deviation or insufficient mechanical intensity may lead to eccentricity to which the eddy current and loss on cans are quite sensitive, causing distortion of airgap flux and inefficient output beyond. The main effect of eccentricity includes: (1) Eddy current density on cans will change according to airgap flux that is affected by the degree and position of the eccentricity; (2) With increase of rotation speed and thus the relative movement between cans, stronger turbulent flow of liquid occurs and slight eccentricity will lead to change of friction loss. Both aspects cause additional loss that raises working temperature and affects system stability. The eccentricity as an important aspect will be studied based on this work.

Appendix

A.1 Deduction of Fourier Coefficients of the Winding Function

According to Fig. A1.1, the trapezoidal waveform distribution of a single turn can be represented by two points $A(\lambda_1, \tau_1)$ and $B(\lambda_2, \tau_2)$. Because of the even symmetry, only the right part with a period $p\pi$ needs to be analyzed. The following can be deduced

$$\begin{aligned} a_v &= \frac{2}{p\pi}\int_0^{p\pi} N(\beta)\cos\left(\frac{v}{p}\beta\right)\mathrm{d}\beta \\ b_v &= 0, \quad v = 1,2,3\ldots \end{aligned} \tag{A1.1}$$

a_v can be further written as

$$\begin{aligned} a_v &= \frac{2}{p\pi}\left(\begin{array}{l} \int_0^{\lambda_1}\tau_1\cos\left(\frac{v}{p}\beta\right)\mathrm{d}\beta + \int_{\lambda_2}^{p\pi}\tau_2\cos\left(\frac{v}{p}\beta\right)\mathrm{d}\beta \\ + \int_{\lambda_1}^{\lambda_2}\left(\frac{\tau_1-\tau_2}{\lambda_1-\lambda_2}\beta + \tau_1 - \frac{\tau_1-\tau_2}{\lambda_1-\lambda_2}\lambda_1\right)\cos\left(\frac{v}{p}\beta\right)\mathrm{d}\beta \end{array}\right), \\ v &= 1,2,3\ldots \end{aligned} \tag{A1.2}$$

which is further reduced to

$$\begin{aligned} a_v &= -\frac{2p}{\pi v^2}s'\left(\cos\frac{v}{p}\lambda_1 - \cos\frac{v}{p}\lambda_2\right), s' = \frac{\tau_1-\tau_2}{\lambda_1-\lambda_2}, \\ v &= 1,2,3\ldots \end{aligned} \tag{A1.3}$$

With $s' = -\frac{\tau_p}{d_w\pi}$, $\lambda_1 = \frac{w_{co}}{\tau_p}\frac{\pi}{2} - \frac{d_w}{\tau_p}\frac{\pi}{2}$ and $\lambda_2 = \frac{w_{co}}{\tau_p}\frac{\pi}{2} + \frac{d_w}{\tau_p}\frac{\pi}{2}$, a_v is further written as

Q. Yu et al., *Analysis and Mathematical Models of Canned Electrical Machine Drives*, https://doi.org/10.1007/978-981-13-2745-2

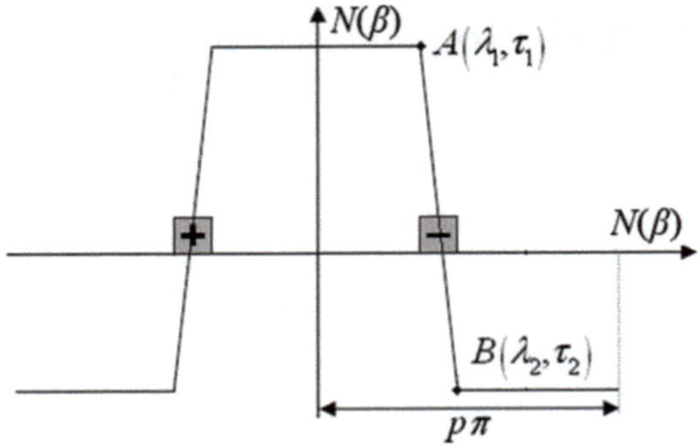

Fig. A1.1 Model of MMF distribution

$$a_v = \frac{2p\tau_p}{\pi^2 v^2 d_w}\left(\cos\left(\frac{v}{p}\left(\frac{w_{co}}{\tau_p}\frac{\pi}{2} - \frac{d_w}{\tau_p}\frac{\pi}{2}\right)\right) - \cos\left(\frac{v}{p}\left(\frac{w_{co}}{\tau_p}\frac{\pi}{2} + \frac{d_w}{\tau_p}\frac{\pi}{2}\right)\right)\right), \quad v = 1, 2, 3\ldots \tag{A1.4}$$

With $\cos(x \pm y) = \cos(x)\cos(y) \mp \sin(x)\sin(y)$, (A1.4) can be further written as

$$\begin{aligned} a_v &= \frac{2p\tau_p}{\pi^2 v^2 d_w}\begin{pmatrix} \cos\left(\frac{v}{p}\frac{w_{co}}{\tau_p}\frac{\pi}{2}\right)\cos\left(\frac{v}{p}\frac{d_w}{\tau_p}\frac{\pi}{2}\right) + \sin\left(\frac{v}{p}\frac{w_{co}}{\tau_p}\frac{\pi}{2}\right)\sin\left(\frac{v}{p}\frac{d_w}{\tau_p}\frac{\pi}{2}\right) \\ -\cos\left(\frac{v}{p}\frac{w_{co}}{\tau_p}\frac{\pi}{2}\right)\cos\left(\frac{v}{p}\frac{d_w}{\tau_p}\frac{\pi}{2}\right) + \sin\left(\frac{v}{p}\frac{w_{co}}{\tau_p}\frac{\pi}{2}\right)\sin\left(\frac{v}{p}\frac{d_w}{\tau_p}\frac{\pi}{2}\right) \end{pmatrix} \\ &= \frac{2p\tau_p}{\pi^2 v^2 d_w}\left(2\sin\left(\frac{v}{p}\frac{w_{co}}{\tau_p}\frac{\pi}{2}\right)\sin\left(\frac{v}{p}\frac{d_w}{\tau_p}\frac{\pi}{2}\right)\right), \\ & v = 1, 2, 3\ldots \end{aligned} \tag{A1.5}$$

Let $\mathrm{si}(x) = \sin(x)/x$, (A1.5) can be finally written as

$$a_v = \frac{2}{\pi v}\left(\sin\left(\frac{v}{p}\frac{w_{co}}{\tau_p}\frac{\pi}{2}\right)\mathrm{si}\left(\frac{v}{p}\frac{d_w}{\tau_p}\frac{\pi}{2}\right)\right), \quad v = 1, 2, 3\ldots \tag{A1.6}$$

A.2 Discretizing Phase Current into Fourier Series

The current waveform can be simplified into a periodical trapezoid characterized by the turn-off angle and current extinguishing angle. Assuming there is N_r number of rotor poles of an SRM and the equivalent pole pair is N_r, phase current alternation speed is N_r times of rotor ω. The Fourier transformation of current can be written as

$$
\begin{aligned}
i &= \frac{a_0}{2} + \sum_{u=1}^{\infty} a_u \cos(uN_r\omega t) + b_u \sin(uN_r\omega t) \\
a_u &= \frac{1}{l}\int_{-l}^{l} i(\omega t)\cos\left(\frac{u\pi\omega t}{l}\right)\mathrm{d}\omega t \\
b_u &= \frac{1}{l}\int_{-l}^{l} i(\omega t)\sin\left(\frac{u\pi\omega t}{l}\right)\mathrm{d}\omega t, \quad u = 1, 2, 3\ldots
\end{aligned}
\tag{A2.1}
$$

where i is the instantaneous phase current that can vary with ωt, u is the current harmonic, a_u and b_u are coefficients, a_0 is the potential part, l is half of period of phase current, $l = \pi/4$. The above equation can be further simplified with even symmetry by

$$
\begin{aligned}
a_u &= \frac{2}{l}\int_{0}^{l} i(\omega t)\cos\left(\frac{u\pi\omega t}{l}\right)\mathrm{d}\omega t \\
b_u &= 0, \quad u = 1, 2, 3\ldots
\end{aligned}
\tag{A2.2}
$$

a_u can be further written as

$$
\begin{aligned}
a_u &= \frac{N_r}{\pi}\int_{0}^{\frac{2\pi}{N_r}} i\cos(N_r u\omega t)\mathrm{d}\omega t = \frac{2N_r}{\pi}\int_{0}^{\frac{\pi}{N_r}} i\cos(N_r u\omega t)\mathrm{d}\omega t \\
&= \frac{2N_r}{\pi}\left(\int_{0}^{\theta_1} I\cos(N_r u\omega t)\mathrm{d}\omega t + \int_{\theta_1}^{\theta_2}\frac{I(\theta_2-\omega t)}{\theta_2-\theta_1}\cos(N_r u\omega t)\mathrm{d}\omega t\right) \\
&= \frac{2I}{\pi N_r(\theta_2-\theta_1)u^2}\left(\cos(N_r\theta_1 u) - \cos(N_r\theta_2 u)\right), \quad u = 1, 2, 3\ldots
\end{aligned}
\tag{A2.3}
$$

$$
a_0 = \frac{2N_r}{\pi}\left(\int_{0}^{\theta_1} I\mathrm{d}\omega t + \int_{\theta_1}^{\theta_2}\frac{I(\theta_2-\omega t)}{\theta_2-\theta_1}\mathrm{d}\omega t\right)
\tag{A2.4}
$$

where I is the peak value of current i. Then the current waveform is written as

$$
i = \frac{a_0}{2} + \sum_{u=1}^{\infty}\frac{2I}{\pi N_r(\theta_2-\theta_1)u^2}\left(\cos(N_r\theta_1 u) - \cos(N_r\theta_2 u)\right)\cos(N_r u\omega t)
\tag{A2.5}
$$

For k-th phase, the current i_k is written as (where N_{ph} is the number of phases)

$$
i_k = \frac{a_0}{2} + 2\sum_{u=1}^{\infty}\frac{I(\cos(N_r u\theta_1) - \cos(N_r u\theta_2))}{\pi N_r(\theta_2-\theta_1)u^2}\cos\left(u\left(N_r\omega t - (k-1)\frac{2\pi}{N_{ph}}\right)\right)
\tag{A2.6}
$$

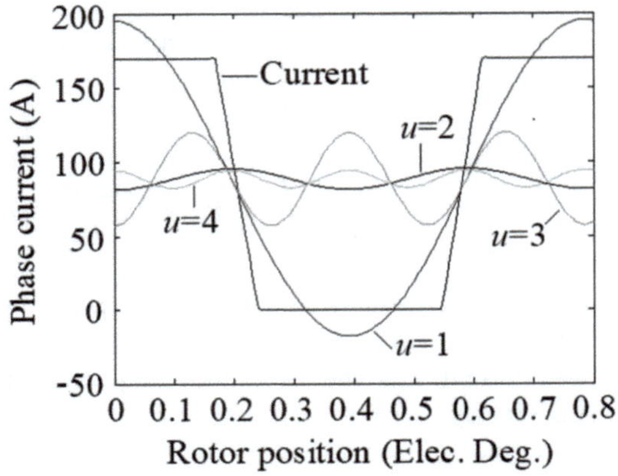

Fig. A2.1 The phase current and harmonics

The turn-off and current extinguishing angles are obtained from the current waveform, regardless of the rotor position. In the following, $\theta_1 = 0.17$ rad, $\theta_2 = 0.24$ rad and $i = 170$A are used. The current harmonics are shown in Fig. A2.1.

A.3 Sum of Coils and Phases

(A) The sum over all coils in one phase leads to

$$\begin{aligned}\sum_{\rho=1}^{2p} e^{j(\rho-1)\pi\left(\frac{v}{p}-1\right)} &= e^{-j\left(\frac{v}{p}-1\right)\pi}\sum_{\rho=1}^{2p} e^{j\rho\left(\frac{v}{p}-1\right)\pi} \\ &= e^{-j\left(\frac{v}{p}-1\right)\pi}\frac{\sin\left(2p\frac{1}{2}\left(\frac{v}{p}-1\right)\pi\right)}{\sin\left(\frac{1}{2}\left(\frac{v}{p}-1\right)\pi\right)}e^{j\frac{1}{2}(2p+1)\left(\frac{v}{p}-1\right)\pi} \\ &= \frac{\sin\left(2p\left(\frac{v}{p}-1\right)\frac{\pi}{2}\right)}{\sin\left(\left(\frac{v}{p}-1\right)\frac{\pi}{2}\right)}e^{j(2p-1)\left(\frac{v}{p}-1\right)\frac{\pi}{2}}, \quad v/p = \pm 1, \pm 2, \pm 3\ldots\end{aligned} \tag{A3.1}$$

Because $p\left(\frac{v}{p}-1\right)$ is an integer number, $\sin\left(2p\left(\frac{v}{p}-1\right)\frac{\pi}{2}\right)=0$ always holds true. Hence, (A3.1) is only different from zero, if

$$\sin\left(\left(\frac{v}{p}-1\right)\frac{\pi}{2}\right)=0 \tag{A3.2}$$

The following relation is obtained

$$\left(\frac{v}{p}-1\right)\frac{\pi}{2}=a\pi, \quad v/p=2a+1, \quad a=0,\pm 1,\pm 2\ldots \tag{A3.3}$$

With the rule of L'Hospital, (A3.1) can be further written as

$$\begin{aligned}\sum_{\rho=1}^{2p} e^{j\left(\frac{v}{p}-1\right)\pi(\rho-1)} &= 2p\frac{\cos\left(2p\left(\frac{v}{p}-1\right)\frac{\pi}{2}\right)}{\cos\left(\left(\frac{v}{p}-1\right)\frac{\pi}{2}\right)}e^{j(2p-1)\left(\frac{v}{p}-1\right)\frac{\pi}{2}}\\ &= 2p\frac{\cos(2pa\pi)}{\cos(a\pi)}e^{j(2p-1)a\pi}\\ &= 2p\frac{\cos(2pa\pi)}{\cos(a\pi)}(\cos(2pa\pi-a\pi)+j\sin(2pa\pi-a\pi))\\ &= 2p\frac{\cos(2pa\pi)\cos(2pa\pi-a\pi)}{\cos(a\pi)}\\ &\frac{v}{p}-1=2a, \quad a=0,\pm 1,\pm 2\ \ldots\end{aligned} \tag{A3.4}$$

With $\cos(x\pm y)=\cos(x)\cos(y)\mp\sin(x)\sin(y)$ used in the term $\cos(2pa\pi-a\pi)$, (A3.4) can be finally simplified as

$$\sum_{\rho=1}^{2p} e^{j\left(\frac{v}{p}-1\right)\pi(\rho-1)}=2p, \quad \frac{v}{p}-1=2a, \quad a=0,\pm 1,\pm 2\ldots \tag{A3.5}$$

(B) The sum over all phases leads to

$$
\begin{aligned}
&\sum_{k=1}^{N_{ph}} e^{j(k-1)\frac{p\pi}{2N_{ph}}\left(\frac{v}{p}-\frac{4}{p}u\right)} \\
&= \sum_{k=1}^{N_{ph}} e^{j(k-1)\frac{p\pi}{2N_{ph}}\left(\frac{v}{p}-2u\right)} = e^{-j\frac{p\pi}{2N_{ph}}\left(\frac{v}{p}-2u\right)} \sum_{k=1}^{N_{ph}} e^{j\frac{p\pi k}{2N_{ph}}\left(\frac{v}{p}-2u\right)} \\
&= e^{-j\frac{p\pi}{2N_{ph}}\left(\frac{v}{p}-2u\right)} \frac{\sin\left(\frac{1}{2}N_{ph}\frac{p\pi}{2N_{ph}}\left(\frac{v}{p}-2u\right)\right)}{\sin\left(\frac{1}{2}\frac{p\pi}{2N_{ph}}\left(\frac{v}{p}-2u\right)\right)} e^{j\frac{1}{2}(N_{ph}+1)\frac{p\pi}{2N_{ph}}\left(\frac{v}{p}-2u\right)} \\
&= \frac{\sin\left(\left(\frac{v}{p}-2u\right)\frac{p}{4}\pi\right)}{\sin\left(\left(\frac{v}{p}-2u\right)\frac{p\pi}{4N_{ph}}\right)} e^{j\frac{1}{2}(N_{ph}-1)\frac{p\pi}{2N_{ph}}\left(\frac{v}{p}-2u\right)}, \\
&\frac{v}{p}-1 = 2a, \quad a = 0, \pm 1, \pm 2\ldots
\end{aligned} \tag{A3.6}
$$

For that $\frac{v}{p}$ is an odd number, $\left(\frac{v}{p}-2u\right)\frac{p\pi}{4N_{ph}} \neq a\pi$ and $\sin\left(\left(\frac{v}{p}-2u\right)\frac{p\pi}{4N_{ph}}\right) \neq 0$ always hold true. (A3.6) will not be further simplified and the sum over all phases is

$$
\begin{aligned}
&\sum_{k=1}^{N_{ph}} e^{j(k-1)\frac{p\pi}{2N_{ph}}\left(\frac{v}{p}-2u\right)} \\
&= \frac{\sin\left(\left(\frac{v}{p}-2u\right)\frac{p}{4}\pi\right)}{\sin\left(\left(\frac{v}{p}-2u\right)\frac{p\pi}{4N_{ph}}\right)} e^{j\frac{1}{2}(N_{ph}-1)\frac{p\pi}{2N_{ph}}\left(\frac{v}{p}-2u\right)} \\
&\frac{v}{p}-1 = 2a, \quad a = 0, \pm 1, \pm 2\ldots, \quad u = 1,2,3\ldots
\end{aligned} \tag{A3.7}
$$

A.4 Deduction of Constants

From the condition Eq. (5.44), where complex forms are underlined,

$$
\begin{aligned}
&{}^{u,\varsigma,v}\underline{E}_{z,i}\big|_{r=r_i} = {}^{u,\varsigma,v}\underline{E}_{z,i+1}\big|_{r=r_i} \quad \Rightarrow \quad -\frac{\partial}{\partial t}{}^{u,\varsigma,v}\underline{A}_{z,i}\big|_{r=r_i} \\
&= -\frac{\partial}{\partial t}{}^{u,\varsigma,v}\underline{A}_{z,i+1}\big|_{r=r_i}
\end{aligned} \tag{A4.1}
$$

with vector potential equations of i-th layer

$$
\begin{aligned}
{}^{u,\varsigma,v}\underline{A}_{z,i} = {}^{u,\varsigma,v}\underline{C}_{3+2i}\left(\underline{I}_{v_p}\left({}^{u,\varsigma}\underline{k}_i r\right) + {}^{u,\varsigma,v}\underline{C}_{4+2i}\underline{K}_{v_p}\left({}^{u,\varsigma}\underline{k}_i r\right)\right)\\
{}^{\varsigma}a\cos\left(\varsigma\frac{\pi}{L_z}z\right)e^{j\left({}^{u}\omega t - v_p(\alpha+\beta)+t_u\right)}
\end{aligned}
\quad (A4.2)
$$

$$
\begin{aligned}
{}^{u,\varsigma,v}\underline{A}_{z,i+1} = {}^{u,\varsigma,v}\underline{C}_{3+2(i+1)}\left(\underline{I}_{v_p}\left({}^{u,\varsigma}\underline{k}_{i+1} r\right) + {}^{u,\varsigma,v}\underline{C}_{4+2(i+1)}\underline{K}_{v_p}\left({}^{u,\varsigma}\underline{k}_{i+1} r\right)\right)\\
{}^{\varsigma}a\cos\left(\varsigma\frac{\pi}{L_z}z\right)e^{j\left({}^{u}\omega t - v_p(\alpha+\beta)+t_u\right)}
\end{aligned}
\quad (A4.3)
$$

We get

$$
\begin{aligned}
&-j^{u}\omega\, {}^{u,\varsigma,v}\underline{C}_{3+2i}\left(\underline{I}_{v_p}\left({}^{u,\varsigma}\underline{k}_i r\right) + {}^{u,\varsigma,v}\underline{C}_{4+2i}\underline{K}_{v_p}\left({}^{u,\varsigma}\underline{k}_i r\right)\right)\Big|_{r=r_i} {}^{\varsigma}a\cos\left(\varsigma\frac{\pi}{L_z}z\right)e^{j\left({}^{u}\omega t - v_p(\alpha+\beta)+t_u\right)}\\
&= -j^{u}\omega\, {}^{u,\varsigma,v}\underline{C}_{3+2(i+1)}\left(\underline{I}_{v_p}\left({}^{u,\varsigma}\underline{k}_{i-1} r\right) + {}^{u,\varsigma,v}\underline{C}_{4+2(i+1)}\underline{K}_{v_p}\left({}^{u,\varsigma}\underline{k}_{i+1} r\right)\right)\Big|_{r=r_i}\\
&\quad {}^{\varsigma}a\cos\left(\varsigma\frac{\pi}{L_z}z\right)e^{j\left({}^{u}\omega t - v_p(\alpha+\beta)+t_u\right)}
\end{aligned}
\quad (A4.4)
$$

Further we get

$$
{}^{u,\varsigma,v}\underline{C}_{3+2i} = {}^{u,\varsigma,v}\underline{C}_{3+2(i+1)}\frac{\left(\underline{I}_{v_p}\left({}^{u,\varsigma}\underline{k}_{i+1} r_i\right) + {}^{u,\varsigma,v}\underline{C}_{4+2(i+1)}\underline{K}_{v_p}\left({}^{u,\varsigma}\underline{k}_{i+1} r_i\right)\right)}{\left(\underline{I}_{v_p}\left({}^{u,\varsigma}\underline{k}_i r_i\right) + {}^{u,\varsigma,v}\underline{C}_{4+2i}\underline{K}_{v_p}\left({}^{u,\varsigma}\underline{k}_i r_i\right)\right)}
\quad (A4.5)
$$

From (5.45)

$$
\begin{aligned}
{}^{u,\varsigma,v}\underline{H}_{\alpha,i}\Big|_{r=r_i} = {}^{u,\varsigma,v}\underline{H}_{\alpha,i+1}\Big|_{r=r_i} \quad \Rightarrow \quad -\frac{1}{\mu_i}\frac{\partial}{\partial r}{}^{u,\varsigma,v}\underline{A}_{z,i}\Big|_{r=r_i}\\
= -\frac{1}{\mu_{i+1}}\frac{\partial}{\partial r}{}^{u,\varsigma,v}\underline{A}_{z,i+1}\Big|_{r=r_i}
\end{aligned}
\quad (A4.6)
$$

We get

$$
\begin{aligned}
&-\frac{1}{\mu_i}\frac{\partial}{\partial r}{}^{u,\varsigma,v}\underline{C}_{3+2i}\left[\underline{I}_{v_p}\left({}^{u,\varsigma}\underline{k}_i r\right) + {}^{u,\varsigma,v}\underline{C}_{4+2i}\underline{K}_{v_p}\left({}^{u,\varsigma}\underline{k}_i r\right)\right]_{r=r_i} {}^{\varsigma}a\cos\left(\varsigma\frac{\pi}{L_z}z\right)e^{j\left({}^{u}\omega t - v_p(\alpha+\beta)+t_u\right)}\\
&\quad = -\frac{1}{\mu_{i+1}}\frac{\partial}{\partial r}{}^{u,\varsigma,v}\underline{C}_{3+2(i+1)}\left[\underline{I}_{v_p}\left({}^{u,\varsigma}\underline{k}_{i+1} r\right) + {}^{u,\varsigma,v}\underline{C}_{4+2(i+1)}\underline{K}_{v_p}\left({}^{u,\varsigma}\underline{k}_{i+1} r\right)\right]_{r=r_i}\\
&\quad {}^{\varsigma}a\cos\left(\varsigma\frac{\pi}{L_z}z\right)e^{j\left({}^{u}\omega t - v_p(\alpha+\beta)+t_u\right)}
\end{aligned}
\quad (A4.7)
$$

Further we get

$$ {}^{u,\varsigma,v}\underline{C}_{3+2i} = {}^{u,\varsigma,v}\underline{C}_{3+2(i+1)} \frac{\mu_i}{\mu_{i+1}} \frac{\frac{\partial}{\partial r}\left[\underline{I}_{v_p}\left({}^{u,\varsigma}\underline{k}_{i+1} r\right) + {}^{u,\varsigma,v}\underline{C}_{4+2(i+1)} \underline{K}_{v_p}\left({}^{u,\varsigma}\underline{k}_{i+1} r\right)\right]_{r=r_i}}{\frac{\partial}{\partial r}\left[\underline{I}_{v_p}\left({}^{u,\varsigma}\underline{k}_{i} r\right) + {}^{u,\varsigma,v}\underline{C}_{4+2i} \underline{K}_{v_p}\left({}^{u,\varsigma}\underline{k}_{i} r\right)\right]_{r=r_i}} \tag{A4.8} $$

For $i = i_{\max} - 1$:

$$ {}^{u,\varsigma,v}\underline{C}_{4+2i} = {}^{u,\varsigma,v}\underline{C}_{4+2(i+1)} = 0 \tag{A4.9} $$

Equations (A4.5) and (A4.7) leads to

$$ \begin{aligned} \frac{{}^{u,\varsigma,v}\underline{C}_{3+2i}}{{}^{u,\varsigma,v}\underline{C}_{3+2(i+1)}} &= \frac{\left(\underline{I}_{v_p}\left({}^{u,\varsigma}\underline{k}_{i+1} r_i\right)\right)}{\left(\underline{I}_{v_p}\left({}^{u,\varsigma}\underline{k}_{i} r_i\right) + {}^{u,\varsigma,v}\underline{C}_{4+2i} \underline{K}_{v_p}\left({}^{u,\varsigma}\underline{k}_{i} r_i\right)\right)} \\ &= \frac{\mu_i}{\mu_{i+1}} \frac{\frac{\partial}{\partial r}\left[\underline{I}_{v_p}\left({}^{u,\varsigma}\underline{k}_{i+1} r\right)\right]_{r=r_i}}{\frac{\partial}{\partial r}\left[\underline{I}_{v_p}\left({}^{u,\varsigma}\underline{k}_{i} r\right) + {}^{u,\varsigma,v}\underline{C}_{4+2i} \underline{K}_{v_p}\left({}^{u,\varsigma}\underline{k}_{i} r\right)\right]_{r=r_i}} \end{aligned} \tag{A4.10} $$

The derivative of modified Bessel functions can be written as [60]

$$ \underline{I}_{v_p}(x) = \frac{1}{2}\left(\underline{I}_{v_p-1}(x) + \underline{I}_{v_p+1}(x)\right) \tag{A4.11} $$

$$ \underline{I}_{v_p}(x) = \frac{1}{2}\left(\underline{K}_{v_p-1}(x) + \underline{K}_{v_p+1}(x)\right) \tag{A4.12} $$

Now Eq. (A4.10) is written as

$$ \begin{aligned} &\mu_{i+1}\underline{I}_{v_p}\left({}^{u,\varsigma}\underline{k}_{i+1} r_i\right) \begin{pmatrix} \underline{I}_{v_p-1}\left({}^{u,\varsigma}\underline{k}_{i} r_i\right) + \underline{I}_{v_p+1}\left({}^{u,\varsigma}\underline{k}_{i} r_i\right) \\ -{}^{u,\varsigma,v}\underline{C}_{4+2i}\left(\underline{K}_{v_p-1}\left({}^{u,\varsigma}\underline{k}_{i} r_i\right) + \underline{K}_{v_p+1}\left({}^{u,\varsigma}\underline{k}_{i} r_i\right)\right) \end{pmatrix} \frac{1}{2} {}^{u,\varsigma}\underline{k}_i \\ &= \mu_i\left(\underline{I}_{v_p}\left({}^{u,\varsigma}\underline{k}_{i} r_i\right) + {}^{u,\varsigma,v}\underline{C}_{4+2i} \underline{K}_{v_p}\left({}^{u,\varsigma}\underline{k}_{i} r_i\right)\right)\left(\underline{I}_{v_p-1}\left({}^{u,\varsigma}\underline{k}_{i+1} r_i\right) + \underline{I}_{v_p+1}\left({}^{u,\varsigma}\underline{k}_{i+1} r_i\right)\right)\frac{1}{2} {}^{u,\varsigma}\underline{k}_{i+1} \end{aligned} \tag{A4.13} $$

Finally we get

$$
{}^{u,\varsigma,v}\underline{C}_{4+2i} = -\frac{\frac{\mu_i}{{}^{u,\varsigma}\underline{k}_i}\underline{I}_{v_p}\left({}^{u,\varsigma}\underline{k}_i r_i\right)\left(\underline{I}_{v_p-1}\left({}^{u,\varsigma}\underline{k}_{i+1} r_i\right)+\underline{I}_{v_p+1}\left({}^{u,\varsigma}\underline{k}_{i+1} r_i\right)\right)-\frac{\mu_{i+1}}{{}^{u,\varsigma}\underline{k}_{i+1}}\underline{I}_{v_p}\left({}^{u,\varsigma}\underline{k}_{i+1} r_i\right)\left(\underline{I}_{v_p-1}\left({}^{u,\varsigma}\underline{k}_i r_i\right)+\underline{I}_{v_p+1}\left({}^{u,\varsigma}\underline{k}_i r_i\right)\right)}{\frac{\mu_i}{{}^{u,\varsigma}\underline{k}_i}\underline{K}_{v_p}\left({}^{u,\varsigma}\underline{k}_i r_i\right)\left(\underline{I}_{v_p-1}\left({}^{u,\varsigma}\underline{k}_{i+1} r_i\right)+\underline{I}_{v_p+1}\left({}^{u,\varsigma}\underline{k}_{i+1} r_i\right)\right)+\frac{\mu_{i+1}}{{}^{u,\varsigma}\underline{k}_{i+1}}\underline{I}_{v_p}\left({}^{u,\varsigma}\underline{k}_{i+1} r_i\right)\left(\underline{K}_{v_p-1}\left({}^{u,\varsigma}\underline{k}_i r_i\right)+\underline{K}_{v_p+1}\left({}^{u,\varsigma}\underline{k}_i r_i\right)\right)} \tag{A4.14}
$$

For $i = 1,2,3,\ldots,i_{\max} - 2$, in analogy we get

$$
\frac{\underline{I}_{v_p}\left({}^{u,\varsigma}\underline{k}_{i+1} r_i\right)+{}^{u,\varsigma,v}\underline{C}_{4+2(i+1)}\underline{K}_{v_p}\left({}^{u,\varsigma}\underline{k}_{i+1} r_i\right)}{\underline{I}_{v_p}\left({}^{u,\varsigma}\underline{k}_i r_i\right)+{}^{u,\varsigma,v}\underline{C}_{4+2i}\underline{K}_{v_p}\left({}^{u,\varsigma}\underline{k}_i r_i\right)} = \frac{\mu_i}{\mu_{i+1}}\frac{\frac{\partial}{\partial r}\left[\underline{I}_{v_p}\left({}^{u,\varsigma}\underline{k}_{i+1} r\right)+{}^{u,\varsigma,v}\underline{C}_{4+2(i+1)}\underline{K}_{v_p}\left({}^{u,\varsigma}\underline{k}_{i+1} r\right)\right]_{r=r_i}}{\frac{\partial}{\partial r}\left[\underline{I}_{v_p}\left({}^{u,\varsigma}\underline{k}_i r\right)+{}^{u,\varsigma,v}\underline{C}_{4+2i}\underline{K}_{v_p}\left({}^{u,\varsigma}\underline{k}_i r\right)\right]_{r=r_i}} \tag{A4.15}
$$

Further we get

$$
\begin{aligned}
&{}^{u,\varsigma,v}\underline{C}_{4+2i}\left\{\begin{aligned}&\frac{\partial}{\partial r}\underline{K}_{v_p}\left({}^{u,\varsigma}\underline{k}_i r\right)\Big|_{r=r_i}\left[\underline{I}_{v_p}\left({}^{u,\varsigma}\underline{k}_{i+1} r_i\right)+{}^{u,\varsigma,v}\underline{C}_{4+2(i+1)}\underline{K}_{v_p}\left({}^{u,\varsigma}\underline{k}_{i+1} r_i\right)\right]\\&-\frac{\mu_i}{\mu_{i+1}}\underline{K}_{v_p}\left({}^{u,\varsigma}\underline{k}_i r_i\right)\frac{\partial}{\partial r}\left[\underline{I}_{v_p}\left({}^{u,\varsigma}\underline{k}_{i+1} r\right)+{}^{u,\varsigma,v}\underline{C}_{4+2(i+1)}\underline{K}_{v_p}\left({}^{u,\varsigma}\underline{k}_{i+1} r\right)\right]_{r=r_i}\end{aligned}\right\}\\
&= -\frac{\partial}{\partial r}\underline{I}_{v_p}\left({}^{u,\varsigma}\underline{k}_i r\right)\Big|_{r=r_i}\left[\underline{I}_{v_p}\left({}^{u,\varsigma}\underline{k}_{i+1} r_i\right)+{}^{u,\varsigma,v}\underline{C}_{4+2(i+1)}\underline{K}_{v_p}\left({}^{u,\varsigma}\underline{k}_{i+1} r_i\right)\right]\\
&+\frac{\mu_i}{\mu_{i+1}}\underline{I}_{v_p}\left({}^{u,\varsigma}\underline{k}_i r_i\right)\frac{\partial}{\partial r}\left[\underline{I}_{v_p}\left({}^{u,\varsigma}\underline{k}_{i+1} r\right)+{}^{u,\varsigma,v}\underline{C}_{4+2(i+1)}\underline{K}_{v_p}\left({}^{u,\varsigma}\underline{k}_{i-1} r\right)\right]_{r=r_i}
\end{aligned} \tag{A4.16}
$$

Using the annotation

$$
\begin{aligned}
{}^{u,\varsigma,v}\underline{K}_{A,4+2(i+1)}\left({}^{u,\varsigma}\underline{k}_{i+1} r_i\right) &= \underline{I}_{v_p-1}\left({}^{u,\varsigma}\underline{k}_{i+1} r_i\right)+\underline{I}_{v_p+1}\left({}^{u,\varsigma}\underline{k}_{i+1} r_i\right)\\
&-{}^{u,\varsigma,v}\underline{C}_{4+2(i+1)}\underline{K}_{v_p-1}\left({}^{u,\varsigma}\underline{k}_{i+1} r_i\right)-{}^{u,\varsigma,v}\underline{C}_{4+2(i+1)}\underline{K}_{v_p+1}\left({}^{u,\varsigma}\underline{k}_{i+1} r_i\right)
\end{aligned} \tag{A4.17}
$$

$$
{}^{u,\varsigma,v}\underline{K}_{B,4+2(i+1)}\left({}^{u,\varsigma}\underline{k}_{i+1} r_i\right) = \underline{I}_{v_p}\left({}^{u,\varsigma}\underline{k}_{i+1} r_i\right)+\underline{K}_{v_p}\left({}^{u,\varsigma}\underline{k}_{i-1} r_i\right) \tag{A4.18}
$$

we get

$$
{}^{u,\varsigma,\nu}\underline{C}_{4+2i} = -\frac{\frac{{}^{u,\varsigma}\underline{k}_{i+1}}{\mu_{i+1}}\underline{I}_{\nu_p}\left({}^{u,\varsigma}\underline{k}_i r_i\right){}^{u,\varsigma,\nu}\underline{K}_{\mathrm{A},4+2(i+1)}\left({}^{u,\varsigma}\underline{k}_{i+1} r_i\right) - \frac{{}^{u,\varsigma}\underline{k}_i}{\mu_i}{}^{u,\varsigma,\nu}\underline{K}_{\mathrm{B},4+2(i+1)}\left({}^{u,\varsigma}\underline{k}_{i+1} r_i\right)\left(\underline{I}_{\nu_p-1}\left({}^{u,\varsigma}\underline{k}_i r_i\right) + \underline{I}_{\nu_p+1}\left({}^{u,\varsigma}\underline{k}_i r_i\right)\right)}{\frac{{}^{u,\varsigma}\underline{k}_{i+1}}{\mu_{i+1}}\underline{K}_{\nu_p}\left({}^{u,\varsigma}\underline{k}_i r_i\right){}^{u,\varsigma,\nu}\underline{K}_{\mathrm{A},4+2(i+1)}\left({}^{u,\varsigma}\underline{k}_{i+1} r_i\right) + \frac{{}^{u,\varsigma}\underline{k}_i}{\mu_i}\underline{K}_{\mathrm{B},4+2(i+1)}\left({}^{u,\varsigma}\underline{k}_{i+1} r_i\right)\left(\underline{K}_{\nu_p-1}\left({}^{u,\varsigma}\underline{k}_i r_i\right) + \underline{K}_{\nu_p+1}\left({}^{u,\varsigma}\underline{k}_i r_i\right)\right)} \tag{A4.19}
$$

With the following equation

$$
{}^{u,\varsigma,\nu}\underline{K}_{4+2(i+1)}\left({}^{u,\varsigma}\underline{k}_{i+1} r_i\right) = \frac{{}^{u,\varsigma,\nu}\underline{K}_{\mathrm{B},4+2(i+1)}\left({}^{u,\varsigma}\underline{k}_{i+1} r_i\right){}^{u,\varsigma}\underline{k}_i \mu_{i+1}}{{}^{u,\varsigma,\nu}\underline{K}_{\mathrm{A},4+2(i+1)}\left({}^{u,\varsigma}\underline{k}_{i+1} r_i\right){}^{u,\varsigma}\underline{k}_{i+1}\mu_i} \tag{A4.20}
$$

we get

$$
{}^{u,\varsigma,\nu}\underline{C}_{4+2i} = -\frac{\underline{I}_{\nu_p}\left({}^{u,\varsigma}\underline{k}_i r_i\right) - {}^{u,\varsigma,\nu}\underline{K}_{4+2(i+1)}\left({}^{u,\varsigma}\underline{k}_{i+1} r_i\right)\left(\underline{I}_{\nu_p-1}\left({}^{u,\varsigma}\underline{k}_i r_i\right) + \underline{I}_{\nu_p+1}\left({}^{u,\varsigma}\underline{k}_i r_i\right)\right)}{\underline{K}_{\nu_p}\left({}^{u,\varsigma}\underline{k}_i r_i\right) + {}^{u,\varsigma,\nu}\underline{K}^{u,\varsigma,\nu}{}_{4+2(i+1)}\left({}^{u,\varsigma}\underline{k}_{i+1} r_i\right)\left(\underline{K}_{\nu_p-1}\left({}^{u,\varsigma}\underline{k}_i r_i\right) + \underline{K}_{\nu_p+1}\left({}^{u,\varsigma}\underline{k}_i r_i\right)\right)} \tag{A4.21}
$$

For $i = 1$, according to (5.43), ${}^{u,\varsigma,\nu}\underline{C}_{3\,+\,2i}$ is now written as

$$
{}^{u,\varsigma,\nu}\underline{C}_5 = -j\mu_1\frac{N_{ph}}{\pi r_0}W^{\nu}\xi_W\sqrt{2}I
\left[\tfrac{1}{2}{}^{u,\varsigma}\underline{k}_1\left(\underline{I}_{\nu_p-1}\left({}^{u,\varsigma}\underline{k}_1 r_0\right) + \underline{I}_{\nu_p+1}\left({}^{u,\varsigma}\underline{k}_1 r_0\right) - {}^{u,\varsigma,\nu}\underline{C}_6\left(\underline{K}_{\nu_p-1}\left({}^{u,\varsigma}\underline{k}_1 r_0\right) + \underline{K}_{\nu_p+1}\left({}^{u,\varsigma}\underline{k}_1 r_0\right)\right)\right)\right]^{-1} \tag{A4.22}
$$

Further (A4.22) is written as

$$
{}^{u,\varsigma,\nu}\underline{C}_{3+2i} = -j\mu_i\frac{N_{ph}}{\pi r_{i-1}}W^{\nu}\xi_W\sqrt{2}I
\frac{2\frac{\mu_i}{{}^{u,\varsigma}\underline{k}_i}}{{}^{u,\varsigma}\underline{k}_i\left(\underline{I}_{\nu_p-1}\left({}^{u,\varsigma}\underline{k}_i r_{i-1}\right) + \underline{I}_{\nu_p+1}\left({}^{u,\varsigma}\underline{k}_i r_{i-1}\right) - {}^{u,\varsigma,\nu}\underline{C}_{4+2i}\left(\underline{K}_{\nu_p-1}\left({}^{u,\varsigma}\underline{k}_i r_{i-1}\right) + \underline{K}_{\nu_p+1}\left({}^{u,\varsigma}\underline{k}_i r_{i-1}\right)\right)\right)} \tag{A4.23}
$$

For $i = 1,2,3,\ldots,i_{\max} - 2$, according to (A4.5)

$$
{}^{u,\varsigma,\nu}\underline{C}_{3+2(i+1)} = {}^{u,\varsigma,\nu}\underline{C}_{3+2i}\frac{\left(\underline{I}_{\nu_p}\left({}^{u,\varsigma}\underline{k}_i r_i\right) + {}^{u,\varsigma,\nu}\underline{C}_{4+2i}\underline{K}_{\nu_p}\left({}^{u,\varsigma}\underline{k}_i r_i\right)\right)}{\left(\underline{I}_{\nu_p}\left({}^{u,\varsigma}\underline{k}_{i+1} r_i\right) + {}^{u,\varsigma,\nu}\underline{C}_{4+2(i+1)}\underline{K}_{\nu_p}\left({}^{u,\varsigma}\underline{k}_{i+1} r_i\right)\right)} \tag{A4.24}
$$

For $i = i_{\max} - 1$ and ${}^{u,\varsigma,\nu}\underline{C}_{4+2i_{\max}} = 0$,

$$ {}^{u,\varsigma,v}\underline{C}_{3+2(i+1)} = {}^{u,\varsigma,v}\underline{C}_{3+2i}\frac{\underline{I}_{v_p}\left({}^{u,\varsigma}\underline{k}_i r_i\right) + {}^{u,\varsigma,v}\underline{C}_{4+2i}\underline{K}_{v_p}\left({}^{u,\varsigma}\underline{k}_i r_i\right)}{\underline{I}_{v_p}\left({}^{u,\varsigma}\underline{k}_{i+1} r_i\right)} \tag{A4.25} $$

A.5 Calculation of Power

According to Poynting's theorem of energy conversion, the generation of power of i-th layer can be further written as

$$ P_{r,i} = -\frac{1}{2}\sum_u\sum_\varsigma\sum_v\left[\begin{array}{l} \left.\int\limits_{-\frac{L_z}{2}}^{\frac{L_z}{2}}\int\limits_{-\pi}^{\pi}\mathrm{Re}\left\{{}^{u,\varsigma,v}\underline{E}_{\alpha,i}{}^{u,\varsigma,v}\underline{H}^*_{z,i} - {}^{u,\varsigma,v}\underline{E}_{z,i}{}^{u,\varsigma,v}\underline{H}^*_{\alpha,i}\right\} r\mathrm{d}\alpha\mathrm{d}z\right|_{r=r_{i-1}} \\ \left.-\int\limits_{-\frac{L_z}{2}}^{\frac{L_z}{2}}\int\limits_{-\pi}^{\pi}\mathrm{Re}\left\{{}^{u,\varsigma,v}\underline{E}_{\alpha,i}{}^{u,\varsigma,v}\underline{H}^*_{z,i} - {}^{u,\varsigma,v}\underline{E}_{z,i}{}^{u,\varsigma,v}\underline{H}^*_{\alpha,i}\right\} r\mathrm{d}\alpha\mathrm{d}z\right|_{r=r_i} \end{array}\right] \tag{A5.1} $$

In (A5.1), components are further deduced as

$$ {}^{u,\varsigma,v}\underline{E}_{\alpha,i} = -\frac{\partial}{\partial t}{}^{u,\varsigma,v}\underline{A}_{\alpha,i} = -j\omega^{u,\varsigma,v}\underline{A}_{\alpha,i} \tag{A5.2} $$

$$ {}^{u,\varsigma,v}\underline{E}_{z,i} = -\frac{\partial}{\partial t}{}^{u,\varsigma,v}\underline{A}_{z,i} = -j\omega^{u,\varsigma,v}\underline{A}_{z,i} \tag{A5.3} $$

$$ \begin{aligned} {}^{u,\varsigma,v}\underline{H}^*_{z,i} &= \frac{1}{\mu_i}{}^{u,\varsigma,v}\underline{B}^*_{z,i} = \frac{1}{\mu_i}\left[\mathrm{rot}\left({}^{u,\varsigma,v}\underline{A}\right)\right]^*_{z,i} \\ &= \frac{1}{\mu_i}\frac{1}{r}\left[\frac{\partial}{\partial r}\left(r^{u,\varsigma,v}\underline{A}_{\alpha,i}\right)\right]^* \\ &= \frac{1}{\mu_i}\frac{1}{r}\left[{}^{u,\varsigma,v}\underline{A}_{\alpha,i} + r\frac{\partial}{\partial r}\left({}^{u,\varsigma,v}\underline{A}_{\alpha,i}\right)\right]^* \end{aligned} \tag{A5.4} $$

$$ \begin{aligned} {}^{u,\varsigma,v}\underline{H}^*_{\alpha,i} &= \frac{1}{\mu_i}{}^{u,\varsigma,v}\underline{B}^*_{\alpha,i} = \frac{1}{\mu_i}\left[\mathrm{rot}\left({}^{u,\varsigma,v}\underline{A}\right)\right]^*_{\alpha,i} \\ &= \frac{1}{\mu_i}\left[-\frac{\partial}{\partial r}\left({}^{u,\varsigma,v}\underline{A}_{z,i}\right)\right]^* \end{aligned} \tag{A5.5} $$

Combining (A5.2), (A5.3), (A5.4) and (A5.5) into (A5.1), it is further written as

$$
\begin{aligned}
&\mathrm{Re}\left\{{}^{u,\varsigma,\nu}\underline{E}_{\alpha,i}\,{}^{u,\varsigma,\nu}\underline{H}^{*}_{z,i} - {}^{u,\varsigma,\nu}\underline{E}_{z,i}\,{}^{u,\varsigma,\nu}\underline{H}^{*}_{\alpha,i}\right\} \\
&= \mathrm{Re}\left\{ -j^{u}\omega^{u,\varsigma,\nu}\underline{A}_{\alpha,i}\frac{1}{\mu_i}\frac{1}{r}\left[{}^{u,\varsigma,\nu}\underline{A}_{\alpha,i} + r\frac{\partial}{\partial r}\left({}^{u,\varsigma,\nu}\underline{A}_{\alpha,i}\right)\right]^{*} + j^{u}\omega^{u,\varsigma,\nu}\underline{A}_{z,i}\frac{1}{\mu_i}\left[-\frac{\partial}{\partial r}\left({}^{u,\varsigma,\nu}\underline{A}_{z,i}\right)\right]^{*}\right\} \\
&= \mathrm{Re}\left\{ -j\frac{{}^{u}\omega}{\mu_i}\frac{1}{r}{}^{u,\varsigma,\nu}\underline{A}_{\alpha,i}\,{}^{u,\varsigma,\nu}\underline{A}^{*}_{\alpha,i} - j\frac{{}^{u}\omega}{\mu_i}{}^{u,\varsigma,\nu}\underline{A}_{\alpha,i}\frac{\partial}{\partial r}\left({}^{u,\varsigma,\nu}\underline{A}_{\alpha,i}\right)^{*} - j\frac{{}^{u}\omega}{\mu_i}{}^{u,\varsigma,\nu}\underline{A}_{z,i}\frac{\partial}{\partial r}\left({}^{u,\varsigma,\nu}\underline{A}_{z,i}\right)^{*}\right\} \\
&= \frac{{}^{u}\omega}{\mu_i}\mathrm{Im}\left\{{}^{u,\varsigma,\nu}\underline{A}_{\alpha,i}\frac{\partial}{\partial r}\left({}^{u,\varsigma,\nu}\underline{A}_{\alpha,i}\right)^{*} + {}^{u,\varsigma,\nu}\underline{A}_{z,i}\frac{\partial}{\partial r}\left({}^{u,\varsigma,\nu}\underline{A}_{z,i}\right)^{*}\right\}
\end{aligned}
\tag{A5.6}
$$

Using the solution in stator fixed coordinate, we get

$$
\begin{aligned}
&\mathrm{Re}\left\{{}^{u,\varsigma,\nu}\underline{E}_{\alpha,i}\,{}^{u,\varsigma,\nu}\underline{H}^{*}_{z,i} - {}^{u,\varsigma,\nu}\underline{E}_{z,i}\,{}^{u,\varsigma,\nu}\underline{H}^{*}_{\alpha,i}\right\} \\
&= \frac{\omega}{\mu_i}\mathrm{Im}\left\{
\begin{array}{l}
\frac{1}{r}{}^{u,\varsigma,\nu}\underline{A}_{\alpha,i}\,{}^{u,\varsigma,\nu}\underline{A}^{*}_{\alpha,i} + {}^{u,\varsigma,\nu}\underline{A}_{\alpha,i}e^{-j\frac{\pi}{2}}\frac{r}{\nu}\frac{\varsigma\pi}{L_z}{}^{u,\varsigma,\nu}\underline{C}^{*}_{3+2i}\frac{1}{2}{}^{u,\varsigma}\underline{k}^{*}_{i}\,{}^{\varsigma}a \\
\left[\begin{array}{l}\underline{I}_{\nu_p}\left({}^{u,\varsigma}\underline{k}_i r\right) + {}^{u,\varsigma,\nu}\underline{C}_{4+2i}\underline{K}_{\nu_p}\left({}^{u,\varsigma}\underline{k}_i r\right) \\ -{}^{u,\varsigma,\nu}\underline{C}_{4+2i}\left[\underline{K}_{\nu_p-1}\left({}^{u,\varsigma}\underline{k}_i r\right) + \underline{K}_{\nu_p+1}\left({}^{u,\varsigma}\underline{k}_i r\right)\right]\end{array}\right]^{*}\sin^2\left(\varsigma\frac{\pi}{L_z}z\right)e^{-j\left({}^{u}\omega t-\nu_p(\alpha+\beta)+t_u\right)} \\
+{}^{u,\varsigma,\nu}\underline{A}_{z,i}\,{}^{u,\varsigma,\nu}\underline{C}^{*}_{3+2i}\frac{1}{2}{}^{u,\varsigma}\underline{k}^{*}_{i}\,{}^{\varsigma}a\cos^2\left(\varsigma\frac{\pi}{L_z}z\right)e^{-j\left({}^{u}\omega t-\nu_p(\alpha+\beta)+t_u\right)} \\
\left[\underline{I}_{\nu_p}\left({}^{u,\varsigma}\underline{k}_i r\right) + {}^{u,\varsigma,\nu}\underline{C}_{4+2i}\underline{K}_{\nu_p}\left({}^{u,\varsigma}\underline{k}_i r\right) - {}^{u,\varsigma,\nu}\underline{C}_{4+2i}\left[\underline{K}_{\nu_p-1}\left({}^{u,\varsigma}\underline{k}_i r\right) + \underline{K}_{\nu_p+1}\left({}^{u,\varsigma}\underline{k}_i r\right)\right]\right]^{*}
\end{array}
\right\}
\end{aligned}
\tag{A5.7}
$$

With the following equation,

$$
{}^{u,\varsigma,\nu}\underline{A}_{\alpha,i}\,{}^{u,\varsigma,\nu}\underline{A}^{*}_{\alpha,i} = 0 \tag{A5.8}
$$

Equation (A5.7) is further written as

$$
\begin{aligned}
&\mathrm{Re}\left\{{}^{u,\varsigma,\nu}\underline{E}_{\alpha,i}\,{}^{u,\varsigma,\nu}\underline{H}^{*}_{z,i} - {}^{u,\varsigma,\nu}\underline{E}_{z,i}\,{}^{u,\varsigma,\nu}\underline{H}^{*}_{\alpha,i}\right\} \\
&= \frac{{}^{u}\omega}{\mu_i}\mathrm{Im}\left\{
\begin{array}{l}
\left[\left(\frac{\varsigma}{\nu}\frac{\pi r}{L_z}\right)^2\sin^2\left(\varsigma\frac{\pi}{L_z}z\right) + \cos^2\left(\varsigma\frac{\pi}{L_z}z\right)\right]{}^{\varsigma}a^2\left|{}^{u,\varsigma,\nu}\underline{C}_{3+2i}\right|^2 \\
\frac{1}{2}{}^{u,\varsigma}\underline{k}^{*}_{i}\left[\underline{I}_{\nu_p}\left({}^{u,\varsigma}\underline{k}_i r\right) + {}^{u,\varsigma,\nu}\underline{C}_{4+2i}\underline{K}_{\nu_p}\left({}^{u,\varsigma}\underline{k}_i r\right)\right] \\
\left[\begin{array}{l}\underline{I}_{\nu_p-1}\left({}^{u,\varsigma}\underline{k}_i r\right) + \underline{I}_{\nu_p+1}\left({}^{u,\varsigma}\underline{k}_i r\right) \\ -{}^{u,\varsigma,\nu}\underline{C}_{4+2i}\left[\underline{K}_{\nu_p-1}\left({}^{u,\varsigma}\underline{k}_i r\right) + \underline{K}_{\nu_p+1}\left({}^{u,\varsigma}\underline{k}_i r\right)\right]\end{array}\right]^{*}
\end{array}
\right\}
\end{aligned}
\tag{A5.9}
$$

$$= \frac{\omega}{2\mu_i}\left[\left(\frac{\varsigma}{v}\frac{\pi r}{L_z}\right)^2 \sin^2\left(\varsigma\frac{\pi}{L_z}z\right) + \cos^2\left(\varsigma\frac{\pi}{L_z}z\right)\right] {}^{\varsigma}a^2 \left|{}^{u,\varsigma,v}\underline{C}_{3+2i}\right|^2$$
$$\operatorname{Im}\left\{\begin{array}{l} {}^{u,\varsigma}\underline{k}_i^* \left[\underline{I}_{v_p}\left({}^{u,\varsigma}\underline{k}_i r\right) + {}^{u,\varsigma,v}\underline{C}_{4+2i}\underline{K}_{v_p}\left({}^{u,\varsigma}\underline{k}_i r\right)\right] \\ \left[\begin{array}{l} \underline{I}_{v_p-1}\left({}^{u,\varsigma}\underline{k}_i r\right) + \underline{I}_{v_p+1}\left({}^{u,\varsigma}\underline{k}_i r\right) \\ -{}^{u,\varsigma,v}\underline{C}_{4+2i}\left[\underline{K}_{v_p-1}\left({}^{u,\varsigma}\underline{k}_i r\right) + \underline{K}_{v_p+1}\left({}^{u,\varsigma}\underline{k}_i r\right)\right] \end{array}\right]^* \end{array}\right\} \tag{A5.10}$$

With the following integral solutions,

$$\int_{-\pi}^{\pi} r\mathrm{d}\alpha = 2\pi r \tag{A5.11}$$

$$\int_{-\frac{L_z}{2}}^{\frac{L_z}{2}} \cos^2\left(\varsigma\frac{\pi}{L_z}z\right)\mathrm{d}\alpha = \frac{L_z}{2} \tag{A5.12}$$

$$\int_{-\frac{L_z}{2}}^{\frac{L_z}{2}} \sin^2\left(\varsigma\frac{\pi}{L_z}z\right)\mathrm{d}\alpha = \frac{L_z}{2} \tag{A5.13}$$

With sum of all harmonics, (A5.10) is finally written as (for stator fixed coordinate)

For $i = 1, 2, 3, \ldots, i_{\max}-1$:

$$\begin{aligned} P_{i\_ave} = \; & -\frac{\pi L_z}{4\mu_i}\sum_u\sum_\varsigma\sum_v {}^{\varsigma}a^2 \left|{}^{u,\varsigma,v}\underline{C}_{3+2i}\right|^2 \frac{{}^{u}\omega}{v_p^2} \\ & \operatorname{Im}\Big\{ {}^{u,\varsigma}\underline{k}_i^* r_{i-1}^3 \left[\left(\tfrac{\varsigma\pi}{L_z}\right)^2 + \left(\tfrac{v_p}{r_{i-1}}\right)^2\right]\left[\underline{I}_{v_p}\left({}^{u,\varsigma}\underline{k}_i r_{i-1}\right) + {}^{u,\varsigma,v}\underline{C}_{4+2i}\underline{K}_{v_p}\left({}^{u,\varsigma}\underline{k}_i r_{i-1}\right)\right] \\ & \times\left[\underline{I}_{v_p-1}\left({}^{u,\varsigma}\underline{k}_i^* r_{i-1}\right) + \underline{I}_{v_p+1}\left({}^{u,\varsigma}\underline{k}_i^* r_{i-1}\right) - {}^{u,\varsigma,v}\underline{C}_{4+2i}^*\left(\underline{K}_{v_p-1}\left({}^{u,\varsigma}\underline{k}_i^* r_{i-1}\right) + \underline{K}_{v_p+1}\left({}^{u,\varsigma}\underline{k}_i^* r_{i-1}\right)\right)\right] \\ & -{}^{u,\varsigma}\underline{k}_i^* r_i^3 \left[\left(\tfrac{\varsigma\pi}{L_z}\right)^2 + \left(\tfrac{v_p}{r_i}\right)^2\right]\left[\underline{I}_{v_p}\left({}^{u,\varsigma}\underline{k}_i r_i\right) + {}^{u,\varsigma,v}\underline{C}_{4+2i}\underline{K}_{v_p}\left({}^{u,\varsigma}\underline{k}_i r_i\right)\right] \\ & \times\left[\underline{I}_{v_p-1}\left({}^{u,\varsigma}\underline{k}_i^* r_i\right) + \underline{I}_{v_p+1}\left({}^{u,\varsigma}\underline{k}_i^* r_i\right) - {}^{u,\varsigma,v}\underline{C}_{4+2i}^*\left(\underline{K}_{v_p-1}\left({}^{u,\varsigma}\underline{k}_i^* r_i\right) - \underline{K}_{v_p+1}\left({}^{u,\varsigma}\underline{k}_i^* r_i\right)\right)\right]\Big\} \end{aligned} \tag{A5.14}$$

For $i = i_{\max}$:

$$P_{i\_ave} = -\frac{\pi L_z}{4\mu_i} \sum_{u} \sum_{\varsigma} \sum_{v} {}^{\varsigma}a^2 \left| {}^{u,\varsigma,v}\underline{C}_{3+2i} \right|^2 \frac{{}^{u}\omega}{v_p^2}$$
$$\mathrm{Im}\left\{ {}^{u,\varsigma}\underline{k}_i^* r_{i-1}^3 \left(\left(\frac{\varsigma\pi}{L_z}\right)^2 + \left(\frac{v_p}{r_{i-1}}\right)^2 \right) \underline{I}_{v_p}\left({}^{u,\varsigma}\underline{k}_i r_{i-1}\right) \left(\underline{I}_{v_p-1}\left({}^{u,\varsigma}\underline{k}_i^* r_{i-1}\right) + \underline{I}_{v_p+1}\left({}^{u,\varsigma}\underline{k}_i^* r_{i-1}\right) \right) \right\} \tag{A5.15}$$

MIX
Papier aus verantwortungsvollen Quellen
Paper from responsible sources
FSC® C105338

If you have any concerns about our products,
you can contact us on
ProductSafety@springernature.com

In case Publisher is established outside the EU,
the EU authorized representative is:
Springer Nature Customer Service Center GmbH
Europaplatz 3, 69115 Heidelberg, Germany

Printed by Libri Plureos GmbH
in Hamburg, Germany